INSECTS IN COLOUR

English Editor
N. D. RILEY

illustrated by
EDGAR HAHNEWALD

BLANDFORD PRESS
POOLE DORSET

First published in the English edition 1963

Link House, West Street,Poole, Dorset BH15 1LL
Second impression 1967
Third impression 1971
Fourth impression 1977
ISBN 0 7137 0144 7

Originally published in Sweden as
Insekter i färg

Printed in Holland by The Ysel Press Ltd, Deventer
and bound by Richard Clay (the Chaucer Press) Ltd. Bungay. Suffolk

CONTENTS

FOREWORD

The coloured illustrations of Insects in this book were drawn by the Swedish artist, Edgar Hahnewald, and were originally published in Sweden and Denmark under the titles *Insekter i Färg* and *Insekter i Farver* respectively. The accompanying descriptions were written by Bengt Olof-Landin (Sweden) and Hans Hvass (Denmark).

A small number of the insects illustrated in the original editions are not known to be present in the British Isles. These have been replaced by new illustrations, supplied by the artist, of other species which are British. It will be found also that some illustrations remain of species which, though common in Scandinavia, are rare or exceedingly rare in Britain. These mostly occur in the extensive coniferous forests of Norway and Sweden and with the extensive planting of such trees in Britain, may be expected to become less rare.

The book is intended to give a summary cross-section of the commonest insects to be found in the countryside, with deliberate emphasis upon the larger kinds, such as butterflies, moths, beetles, dragonflies, wasps, bees, ants and grasshoppers most likely to be met with. Species typical of most of the less obvious but important groups are also illustrated and several uncommon species have also been deliberately included to illustrate some point of especial interest concerning their habits or occurrence. In all 260 species are figured representing about 4 out of every 300 kinds known to occur in the British Isles. In the brief introduction a short general outline is given of the more salient features of insect anatomy, biology and classification.

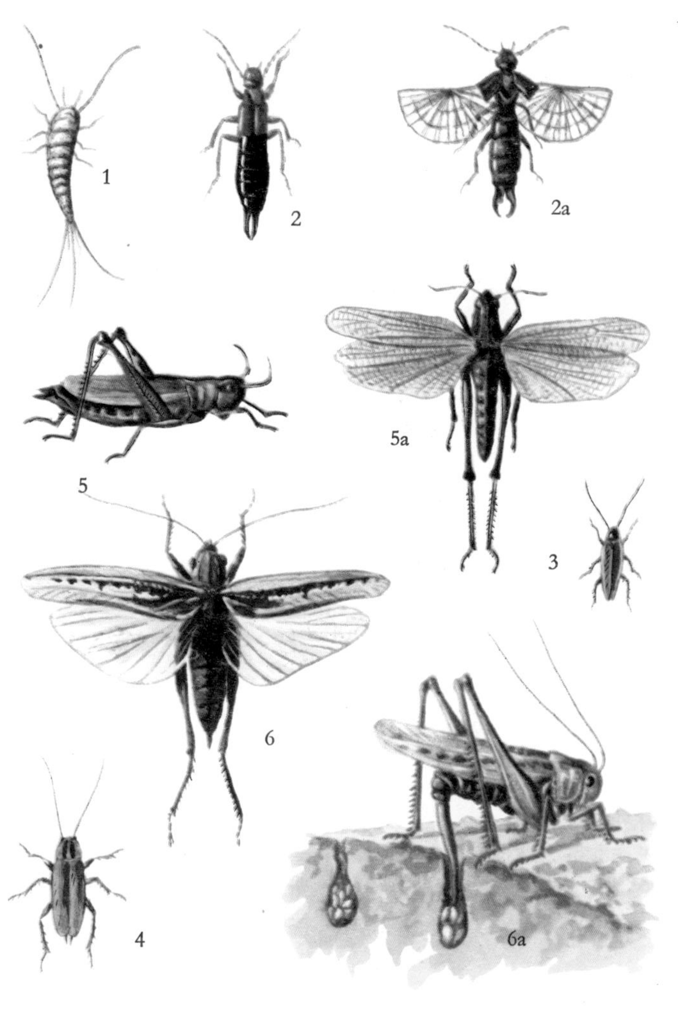

1 **Silver Fish,** *Lepisma saccharina* 2 **Earwig,** *Forficula auricularia,* female 2a male with wings outspread 3 **Lapp Cockroach,** *Ectobius lapponicus* 4 **German Cockroach,** *Blatella germanica* 5 **Common Grasshopper,** *Chorthippus bicolor,* male 5a female with wings open 6 **Wart-biter,** *Decticus verrucivorus,* male with wings open 6a female, egg-laying

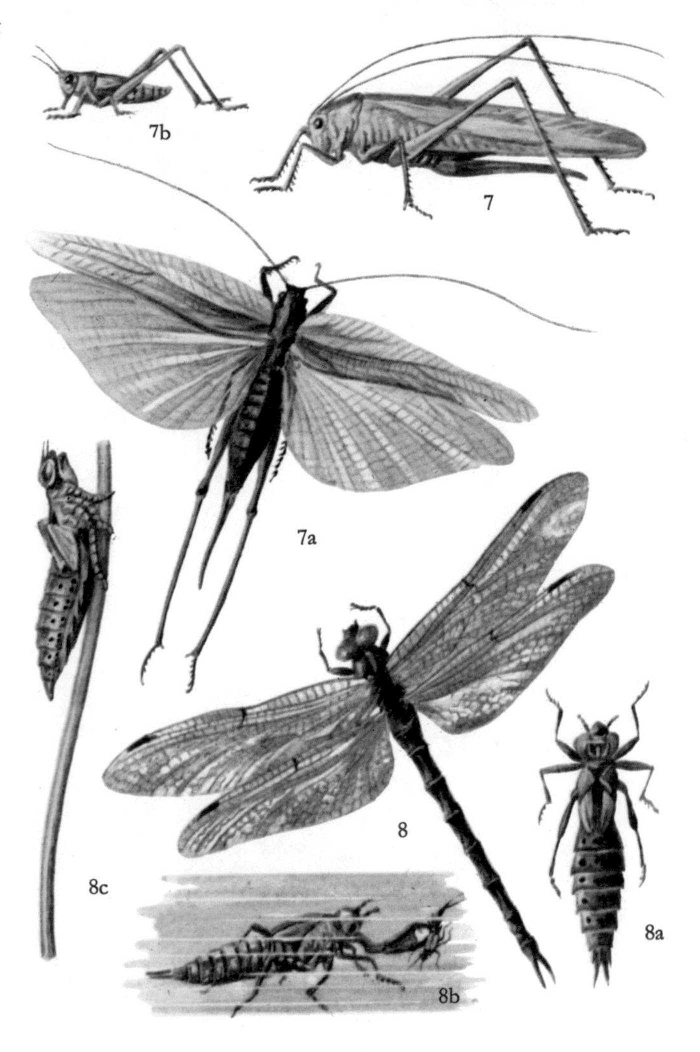

7 **Great Green Grasshopper,** *Tettigonia viridissima,* female 7a female with wings open 7b young hopper
8 **Brown Aeshna,** *Aeshna grandis* 8a nymph 8b nymph showing action of the 'mask' 8c nymphal skin after emergence of the adult dragonfly

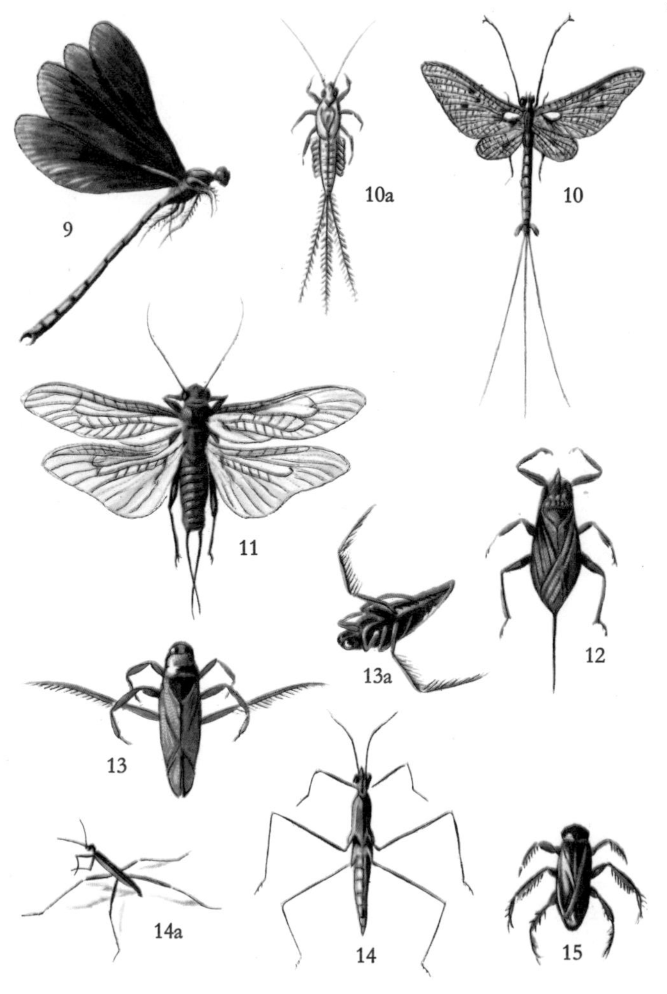

9 **The Demoiselle,** *Agrion virgo,* male 10 **Mayfly,** *Ephemera vulgata* 10a nymph 11 **Stonefly,** *Perla cephalotes* 12 **Water Scorpion,** *Nepa cinerea* 13 **Water Boatman,** *Notonecta glauca* 13a in swimming position, on its back 14 **Great Pond Skater,** *Gerris naias* 14a on the surface of the water 15 **Lesser Water Boatman,** *Corixa punctata*

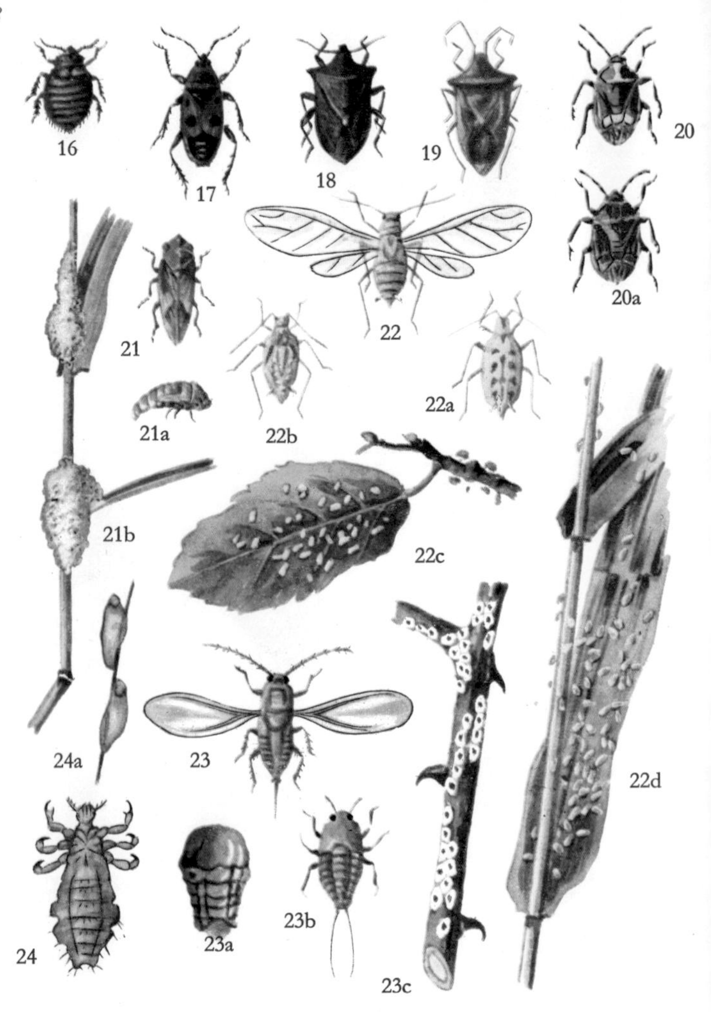

16 **Bed Bug,** *Cimex lectularius* 17 **Beadle or Soldier Bug,** *Pyrrhocoris apterus* 18 **Toothed Shield Bug,** *Picromerus bidens* 19 **Blood-red Shield Bug,** *Acanthosoma haemorrhoidale* 20 **Cabbage Shield Bug,** *Eurydema oleraceum* 20a colour variety 21 **Alder Frog-hopper,** *Aphrophora alni* 21a nymph 21b cuckoo-spit on grass stem 22 **Mealy Plum Aphis,** *Hyalopterus arundinis,* winged form 22a wingless (apterous) form 22b nymph 22c aphids on plum-leaf 22d on rush 23 **Scurvy Scale,** *Aulacaspis rosae,* winged male 23a wingless female 23b nymph 23c scales on rose stem 24 **Head Louse,** *Pediculus humanus* 24a eggs (nits) on hair

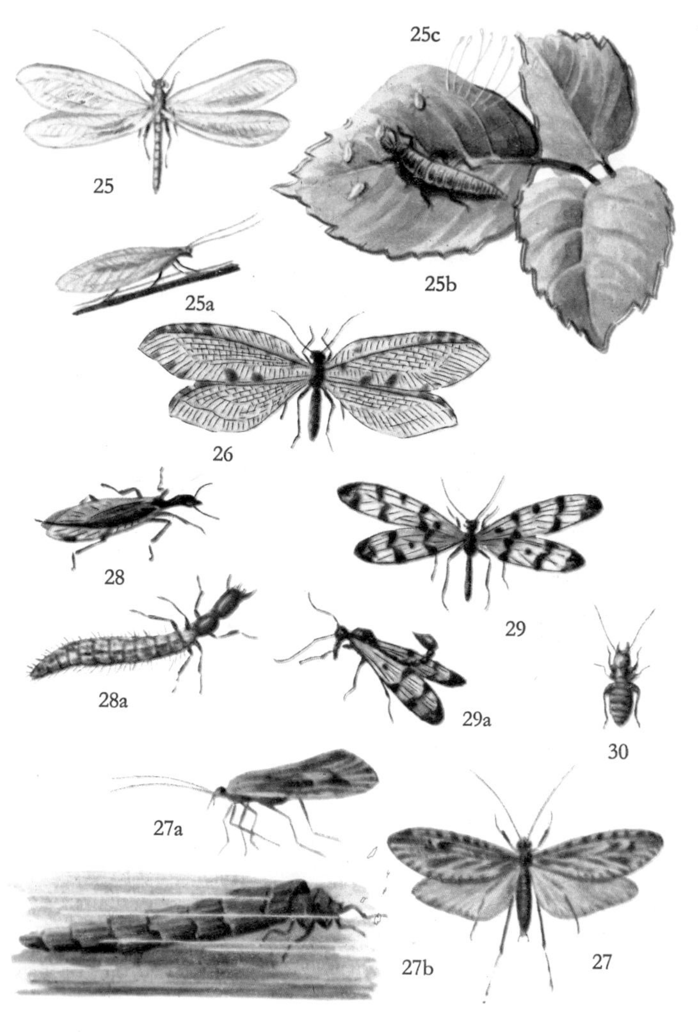

25 **Green Lacewing,** *Chrysopa carnea* 25a in resting position 25b larva 25c stalked eggs 26 **Giant Lacewing,** *Osmylus fulvicephalus* 27 **Great Red Sedge,** *Phryganea grandis* 27a in resting position 27b larva and case 28 **Snake Fly,** *Raphidia notata,* 28a larva 29 **Scorpion Fly,** *Panorpa communis,* 29a in resting position 30 **Book Louse,** *Liposcelis divinatorius*

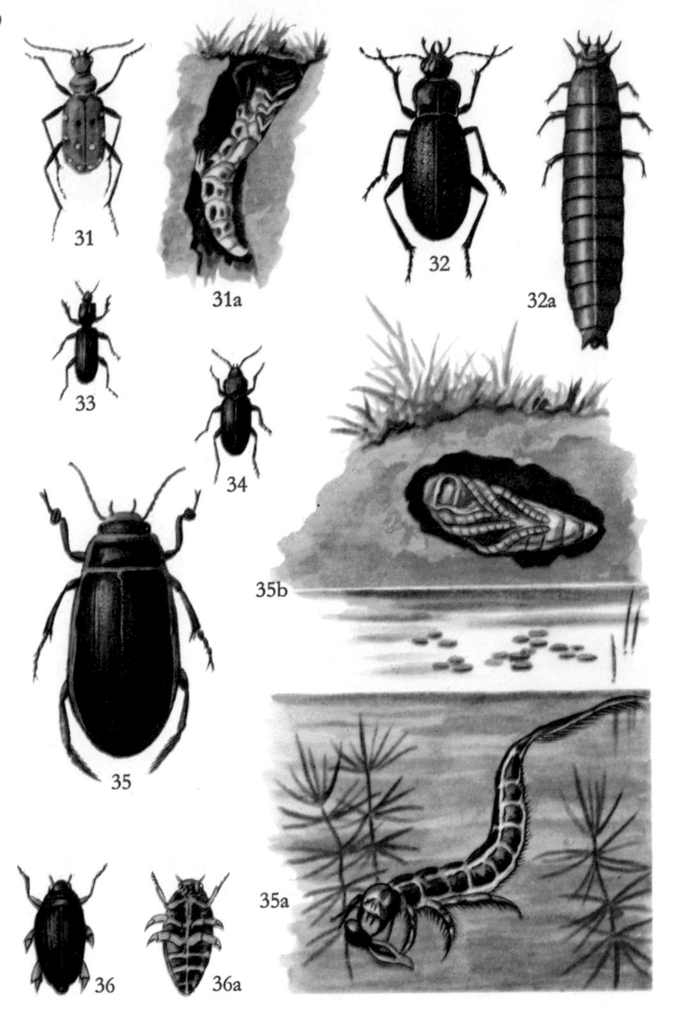

31 **Tiger Beetle,** *Cicindela campestris* 31a larva 32 **Woodland Ground Beetle,** *Carabus nemoralis,* 32a larva 33 **Burrowing Ground Beetle,** *Clivina fossor* 34 **Shiny Ground Beetle,** *Harpalus aeneus* 35 **Giant Water Beetle,** *Dytiscus marginalis,* male 35a larva, 35b pupa 36 **Whirligig,** *Gyrinus substriatus* 36a underside

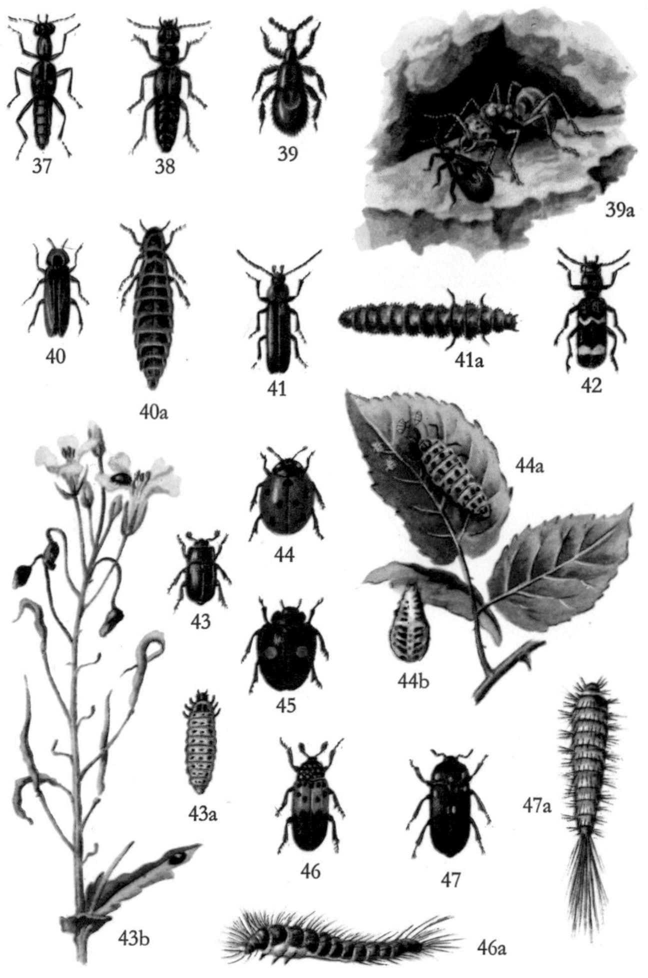

37 **Black Rove Beetle,** *Stenus biguttatus* 38 **Red-backed Rove Beetle** *Staphylinus caesareus* 39 **Ant-nest Beetle,** *Claviger testaceus* 39a beetle being licked by yellow ant 40 **Glow-worm,** *Lampyris noctiluca,* male 40a female 41 **Sailor Beetle,** *Cantharis fusca* 41a larva 42 **Ant Beetle,** *Thanasimus formicarius* 43 **Brassy Pollen Beetle,** *Meligethes aeneus* 43a larva 43b on flowers of rape 44 **Seven-spot Ladybird,** *Coccinella septempunctata* 44a larva 44b pupa 45 **False Ladybird,** *Chilocorus renipustulatus* 46 **Bacon Beetle,** *Dermestes lardarius* 46a larva 47 **Fur Beetle,** *Attagenus pellio* 47a larva

48 **Common Burying Beetle,** *Necrophorus vespillo* 48a larva 48b pupa
49 **Black Burying Beetle,** *Necrophorus humator*
50 **Red-necked Sexton,** *Oeceoptoma thoracicum* 50a larva
51 **Nature's Scavengers,** The same beetles with a dead titmouse

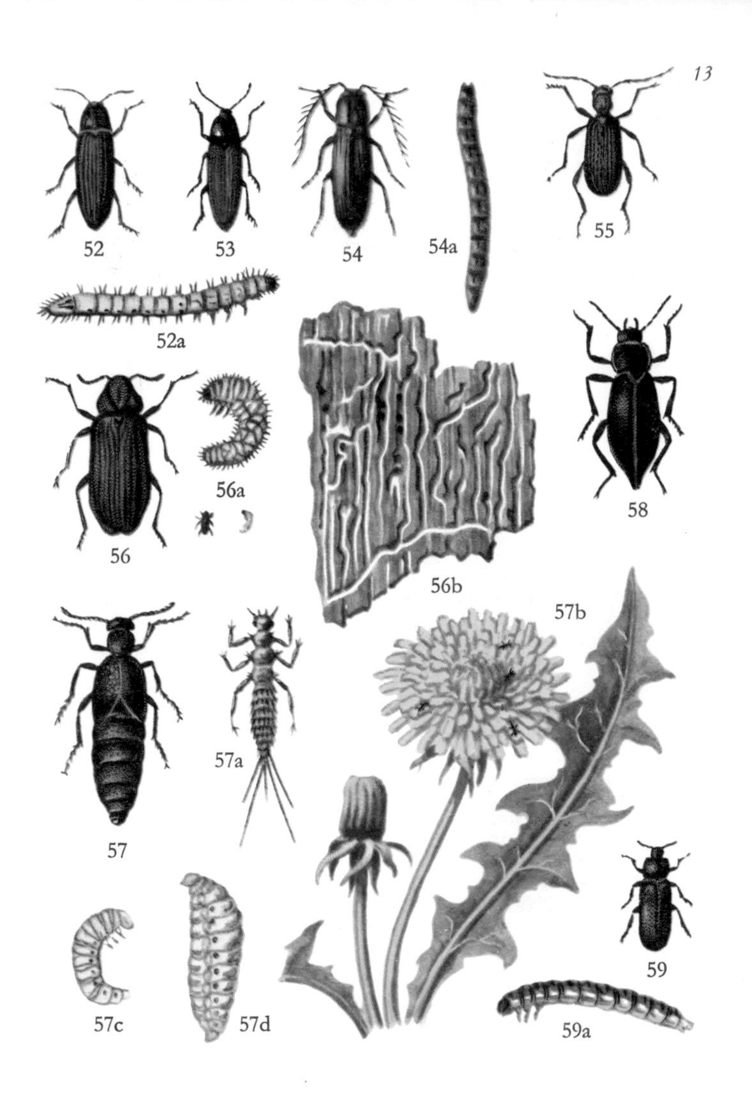

52 **Lined Click-beetle,** *Agriotes lineatus,* 52a larva 53 **Scarlet Click-beetle,** *Elater sanguineus* 54 **Comb-horned Upland Click-beetle,** *Corymbites pectinicornis* 54a larva 55 **Spider Beetle,** *Ptinus fur* 56 **Furniture Beetle,** *Anobium punctatum* 56a larva 56b workings of larva in wood, also beetle and larva natural size 57 **Oil Beetle,** *Meloe proscarabaeus,* 57a larva in first stage (triungulin) 57b triungulin larva on dandelion 57c second stage larva 57d third stage larva 58 **Cellar Beetle,** *Blaps mortisaga* 59 **Mealworm Beetle,** *Tenebrio molitor* 59a Mealworm (larva)

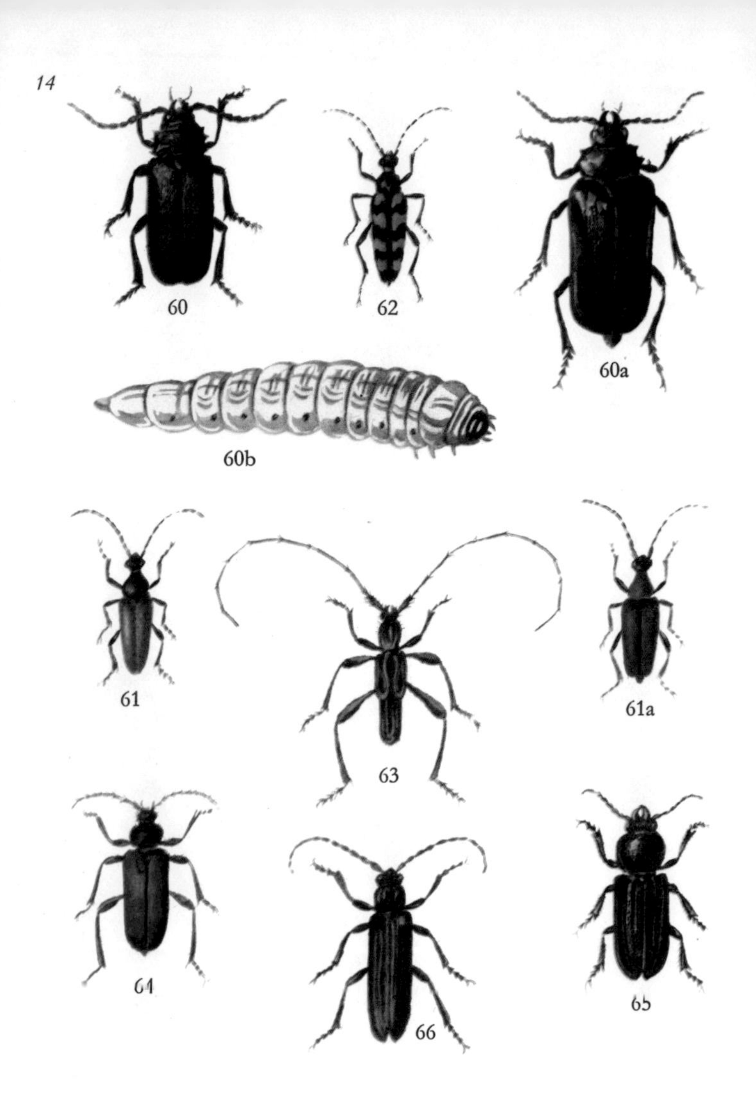

60 **The Tanner,** *Prionus coriarius,* male 60a female 60b larva 61 **Red-brown Long-horn Beetle,** *Leptura rubra,* male 61a female 62 **Four-banded Long-horn Beetle,** *Strangalia quadrifasciata* 63 **Short-winged Long-horn Beetle,** *Molorchus minor* 64 **Violet Long-horn Beetle,** *Callidium violaceum* 65 **Firewood Long-horn Beetle,** *Spondylis buprestoides* 66 **Dusky Long-horn Beetle,** *Criocephalus rusticus*

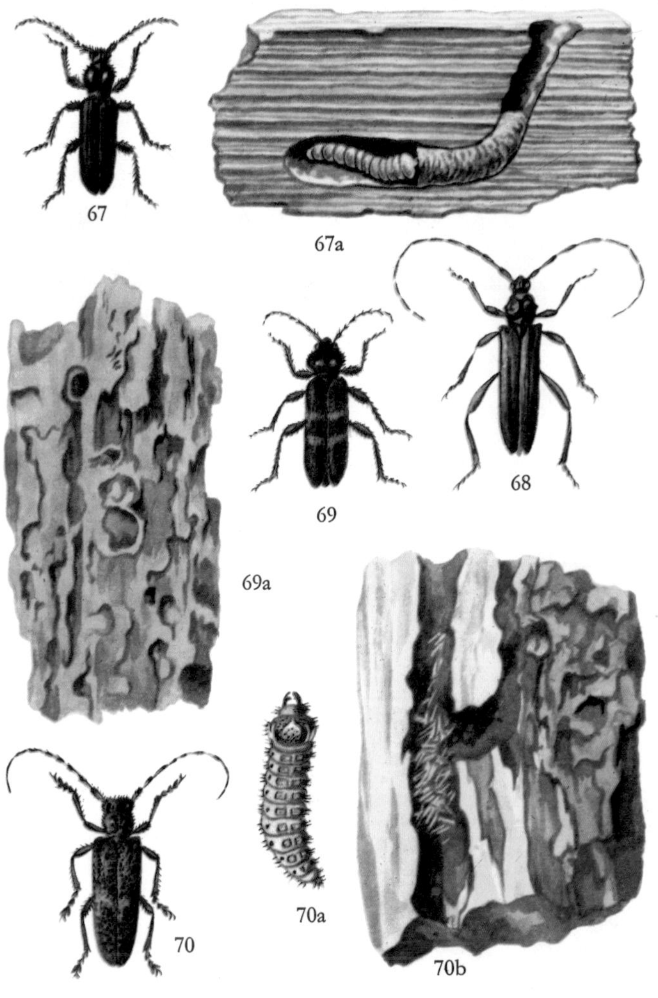

67 **Chestnut Long-horn Beetle,** *Tetropium castaneum* 67a larva in gallery in wood 68 **Musk Beetle,** *Aromia moschata* 69 **House Long-horn Beetle,** *Hylotrupes bajulus* 69a its larva burrowing in wood 70 **Large Poplar Long-horn,** *Saperda carcharias* 70a larva 70b larval galleries in wood of poplar

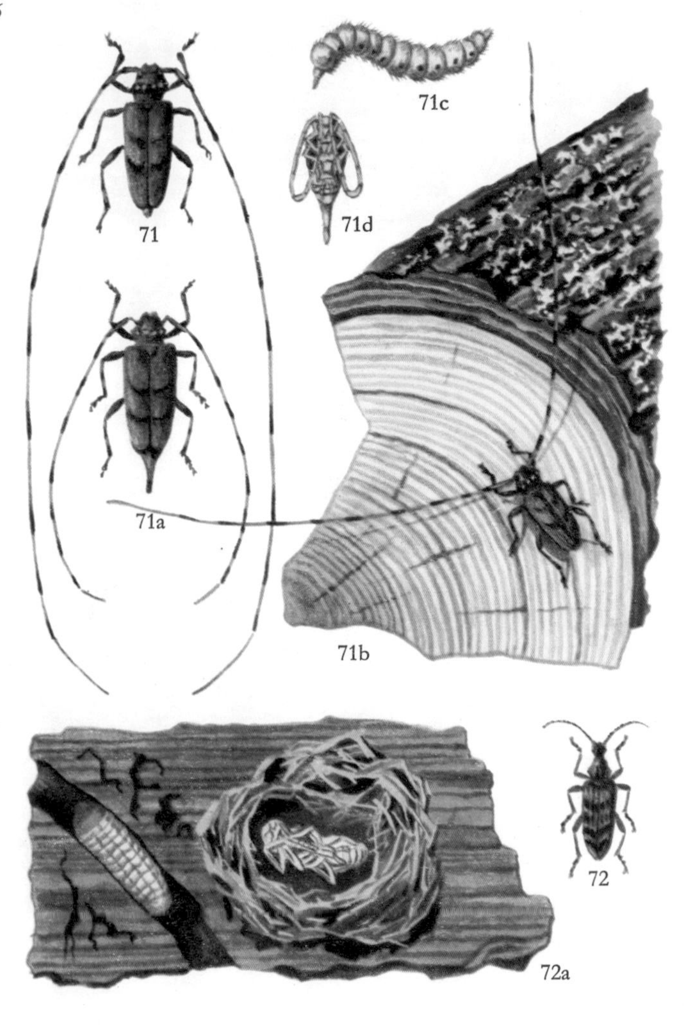

71 **Timberman,** *Acanthocinus aedilis,* male 71a female 71b male on felled log 71c larva, 71d pupa

72 **Pine-bark Long-horn,** *Rhagium inquisitor* 72a larva in gallery and pupa in pupal chamber

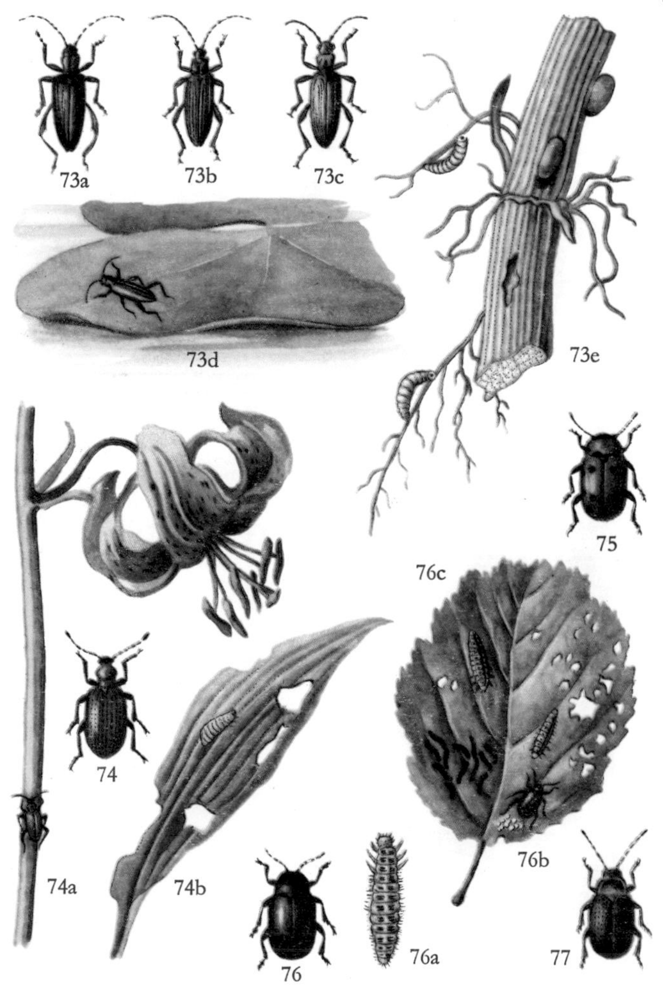

73 **Water Jewels** 73a *Donacia aquatica* 73b *Donacia clavipes* 73c *Donacia cinerea* 73d *Donacia crassipes* 73e larvae and pupae 74 **Scarlet Lily Beetle,** *Lilioceris lilii* 74a beetle on lily stem (natural size) 74b larva 75 **Four-spot Willow Beetle,** *Phytodecta viminalis* 76 **Alder Leaf Beetle,** *Agelastica alni* 76a larva 76b egg-laying beetle and larvae on alder leaf 77 **Sallow Leaf Beetle,** *Lochmaea capreae*

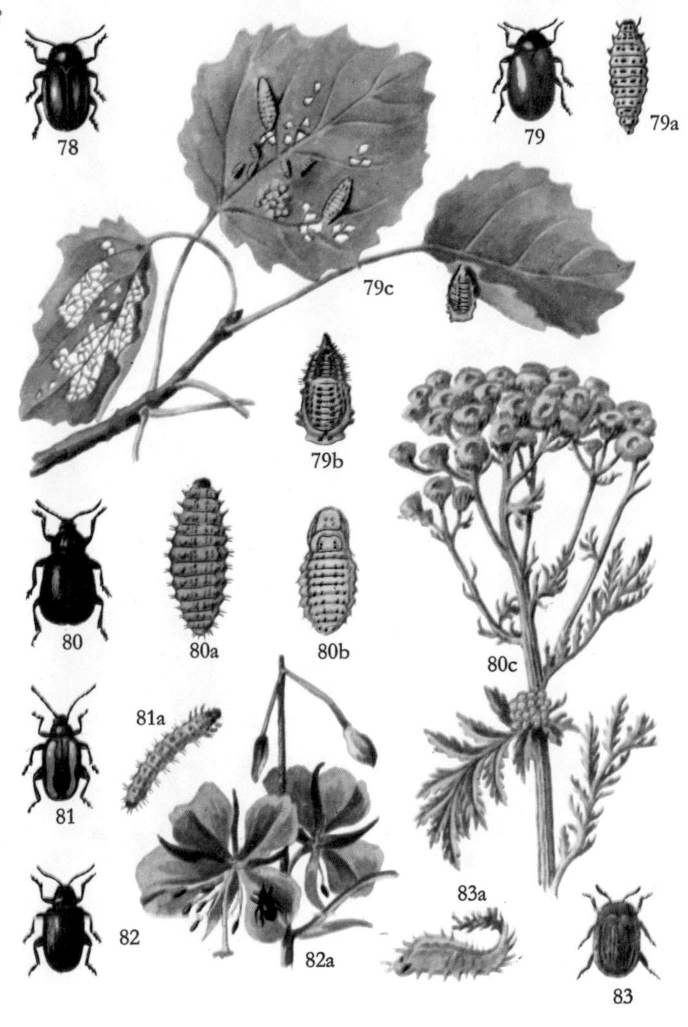

78 **White Deadnettle Beetle,** *Chrysomela fastuosa* 79 **Red Poplar Leaf Beetle,** *Melasoma populi* 79a larva 79b pupa 79c egg cluster, larvae and pupae on poplar 80 **Tansy Leaf Beetle,** *Galeruca tanaceti* 80a larva 80b pupa 80c egg cluster on tansy 81 **Turnip Flea Beetle,** *Phyllotreta nemorum* 81a larva 82 **Cabbage Leaf Beetle,** *Haltica oleracea* 82a beetle on willow-herb 83 **Tortoise Beetle,** *Cassida nebulosa* 83a larva

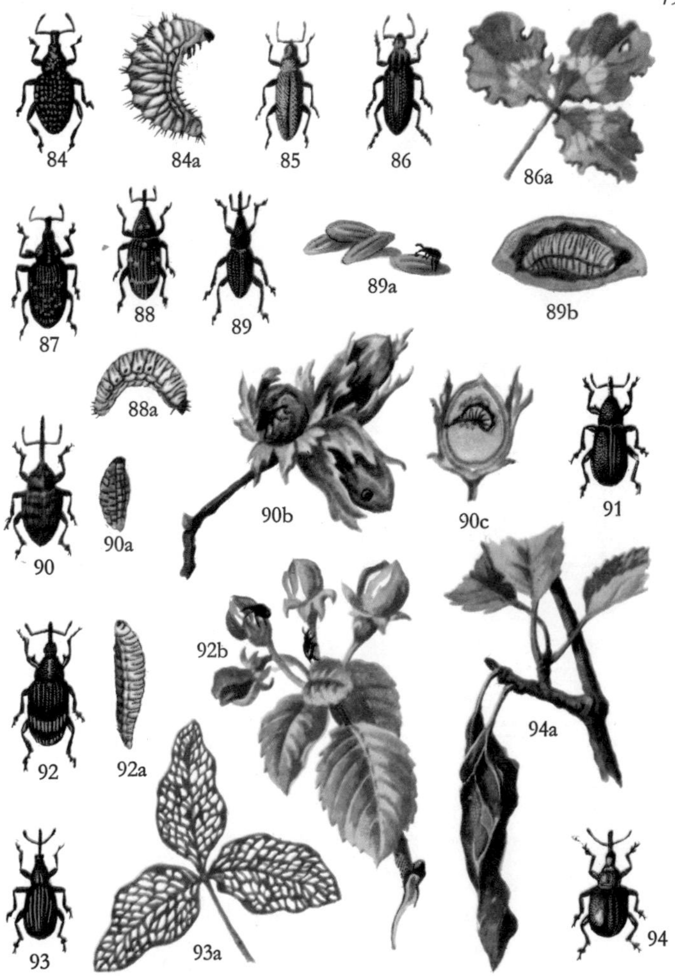

84 **Vine Weevil,** *Ótiorrhynchus sulcatus* 84a larva 85 **Silver-green Leaf Weevil,** *Phyllobius argentatus* 86 **Pea Weevil,** *Sitona lineatus* 86a clover leaf attacked by beetle 87 **Pine Weevil,** *Hylobius abietis* 88 **Banded Pine Weevil,** *Pissodes pini* 88a larva 89 **Grain Weevil,** *Calandra granariae* 89a beetle on corn 89b larva inside grain 90 **Nut Weevil,** *Balaninus nucum* 90a larva 90b beetle and borehole in nut 90c larva inside hazel nut 91 **Dark Blue Pine Weevil,** *Magdalis violacea* 92 **Apple Blossom Weevil,** *Anthonomus pomorum* 92a larva 92b beetles on opening apple blossom 93 **Clover Weevil,** *Apion apricans,* 93a clover leaf skeletonised by beetle 94 **Birch Leaf Roller,** *Bycticus betulae* 94a birch leaves rolled by the beetle

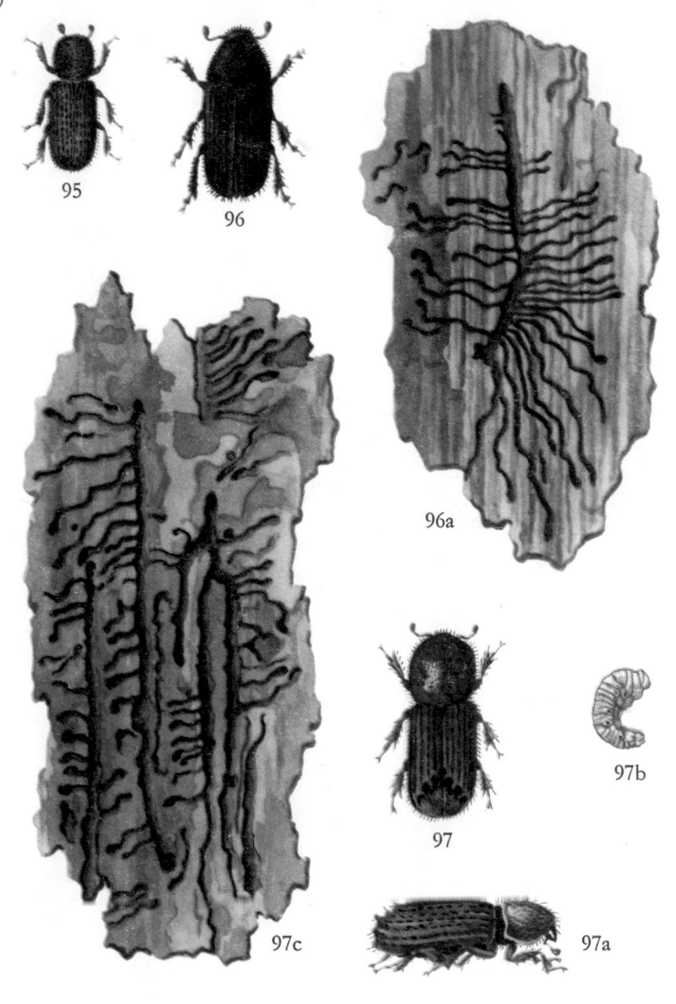

95 **Bark Beetle,** *Crypturgus pusillus* 96 **Pine Shoot Beetle,** *Myelophilus piniperda* 96a borings made by larvae under bark 97 **The Pattern Maker,** *Ips typographus* 97a side view, 97b larva 97c borings made by larvae under bark

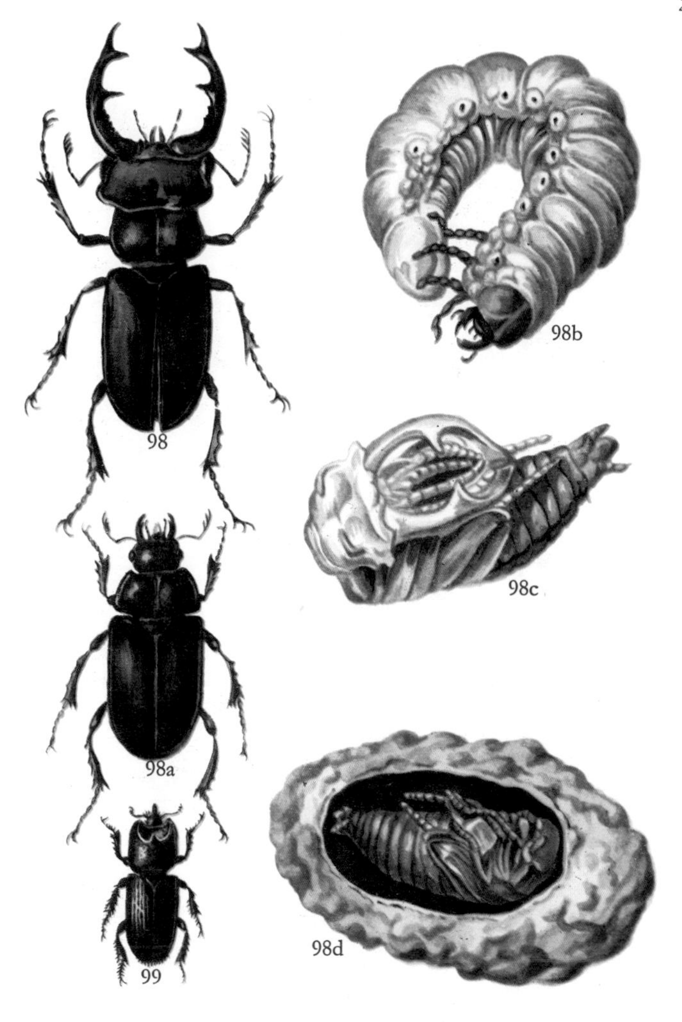

98 **Stag Beetle,** *Lucanus cervus,* male 98a female 98b larva 98c male pupa 98d female pupa in opened cocoon 99 **Lesser Stag Beetle,** *Sinodendron cylindricus*

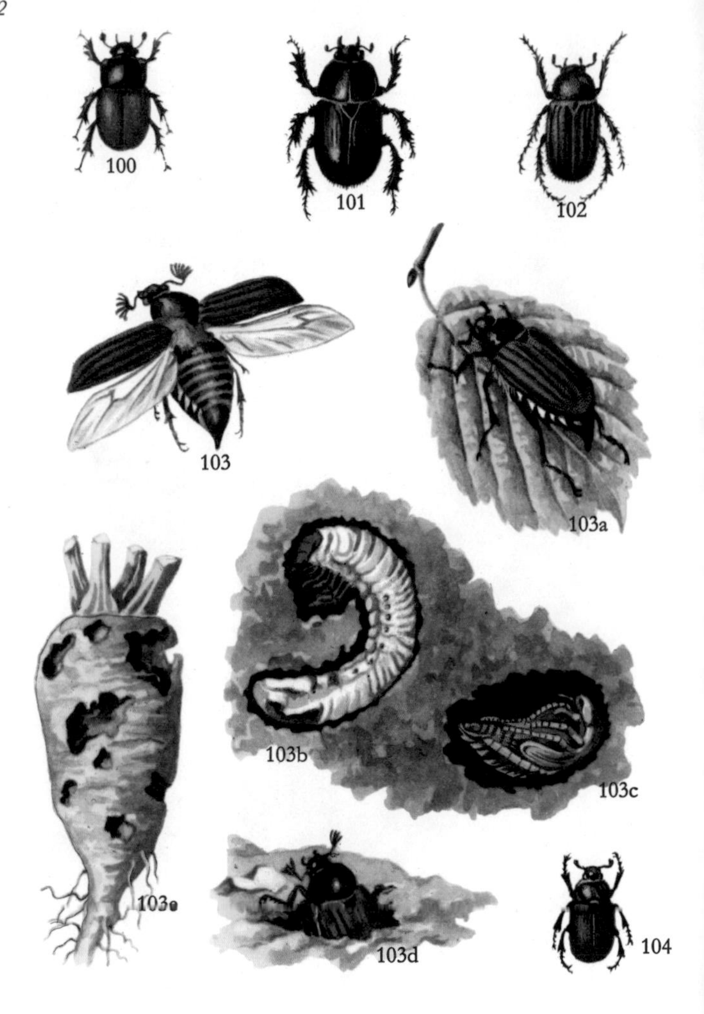

100 **Dung Beetle,** *Aphodius fimetarius* 101 **Dor Beetle,** *Geotrupes stercorarius* 102 **Summer Chafer,** *Amphimallus solstitialis* 103 **Cockchafer,** *Melolontha melolontha,* male 103a female 103b larva 103c pupa 103d newly hatched beetle emerging from soil 103e sugar beet attacked by larvae 104 **Garden Chafer,** *Phyllopertha horticola*

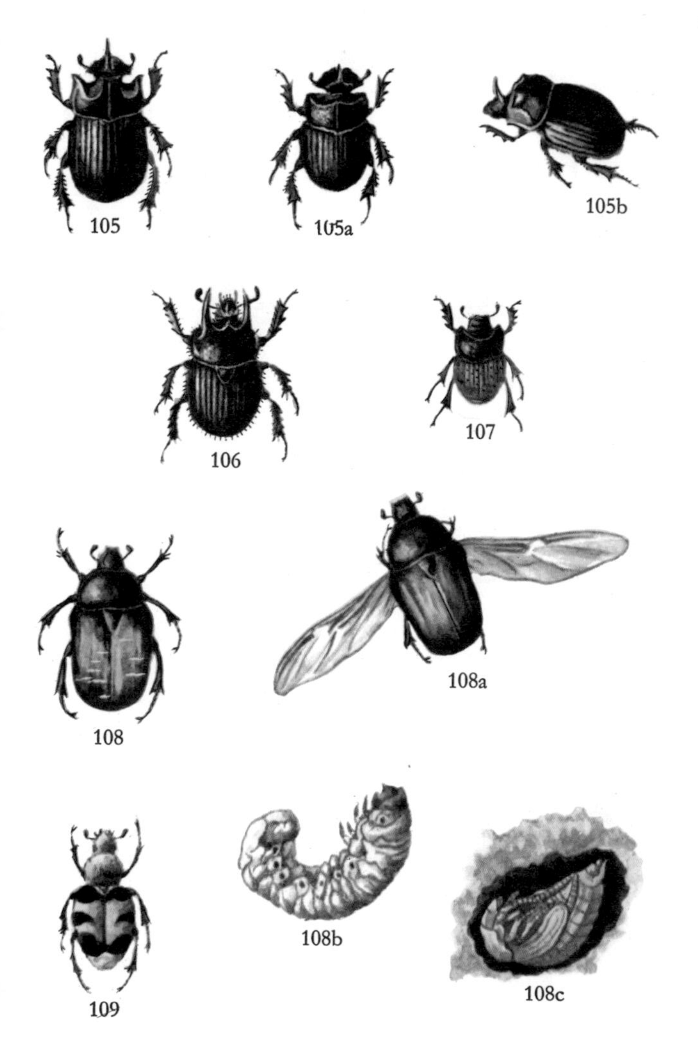

105 **One-horned Dung Beetle,** *Copris lunaris* 105a female 105b male profile 106 **Three-horned Dor Beetle,** *Typhaeus typhaeus* 107 **Bronze Dung Beetle,** *Onthophagus vacca* 108 **Rose Chafer,** *Cetonia aurata* 108a beetle in flight 108b larva 108c pupa 109 **Bee Chafer,** *Trichius fasciatus*

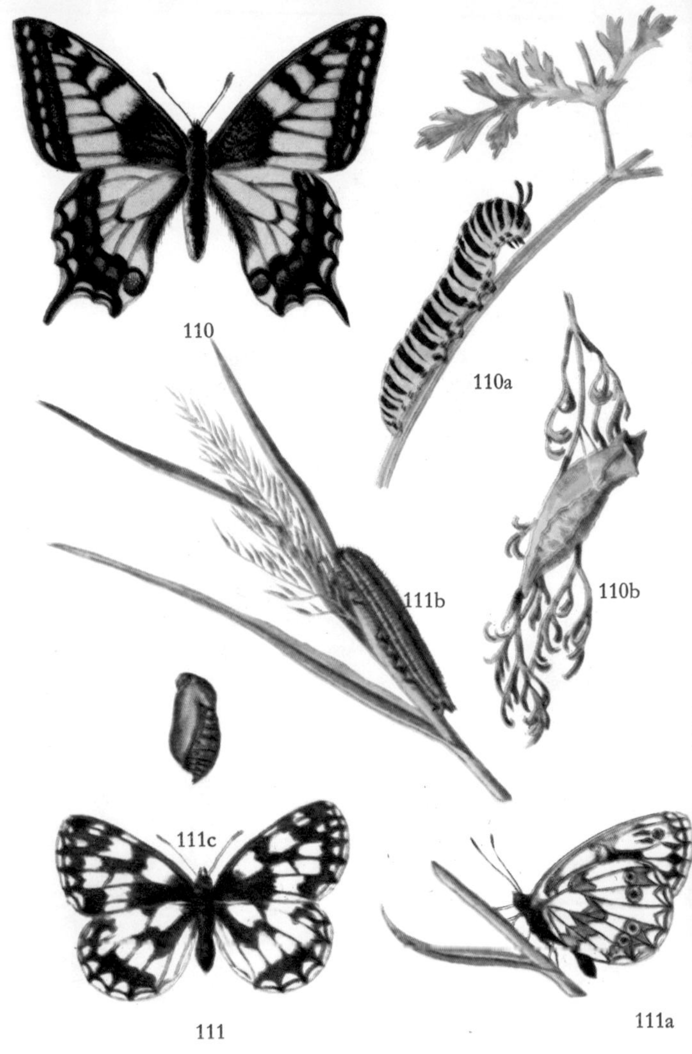

110 **Swallowtail,** *Papilio machaon* 110a larva 110b pupa
111 **Marbled White,** *Melanargia galathea* 111a underside 111b larva 111c pupa

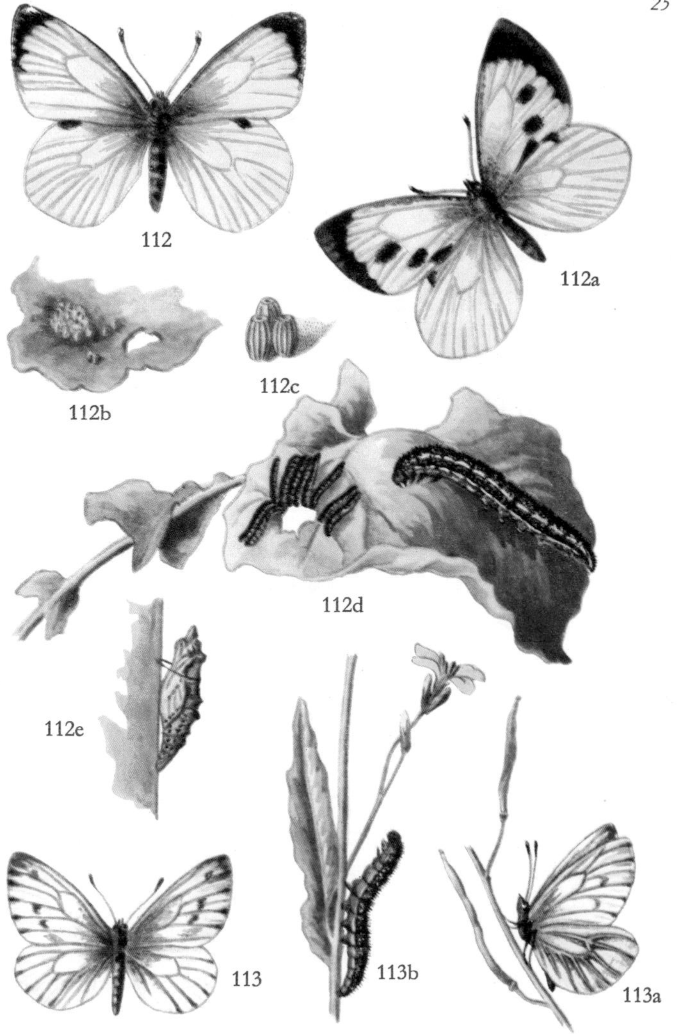

112 **Large White,** *Pieris brassicae,* male 112a female 112b egg cluster
112c eggs (enlarged) 112d caterpillars 112e chrysalis
113 **Green-veined White,** *Pieris napi* 113a under-side 113b caterpillar

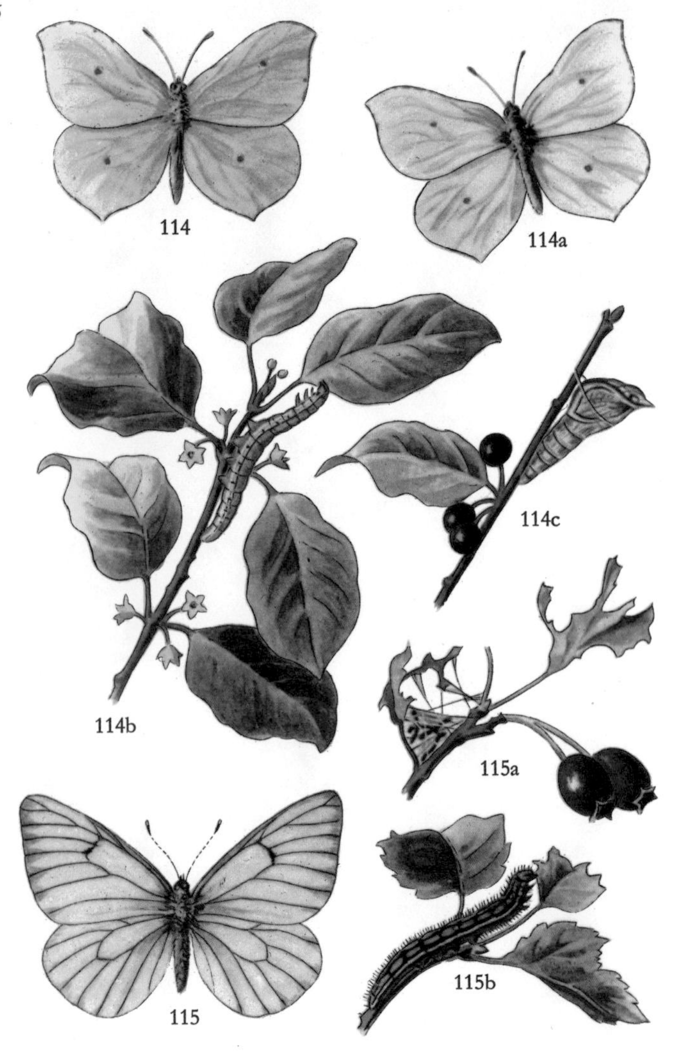

114 **Brimstone,** *Gonepteryx rhamni,* male 114a female 114b caterpillar on buckthorn 114c chrysalis

115 **Black-veined White,** *Aporia crataegi* 115a young caterpillar in web on hawthorn 115b full-grown caterpillar

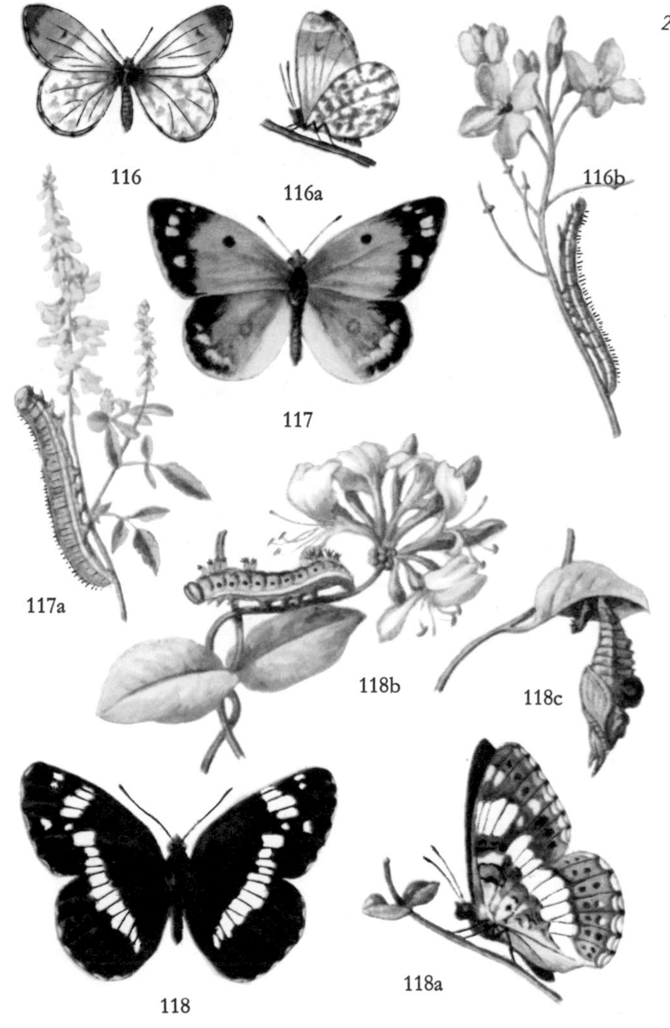

116 **Orange Tip,** *Anthocharis cardamines,* male 116a under-side 116b caterpillar
117 **Clouded Yellow,** *Colias crocea* 117a larva
118 **White Admiral,** *Limenitis camilla* 118a under-side 118b larva on honeysuckle 118c pupa

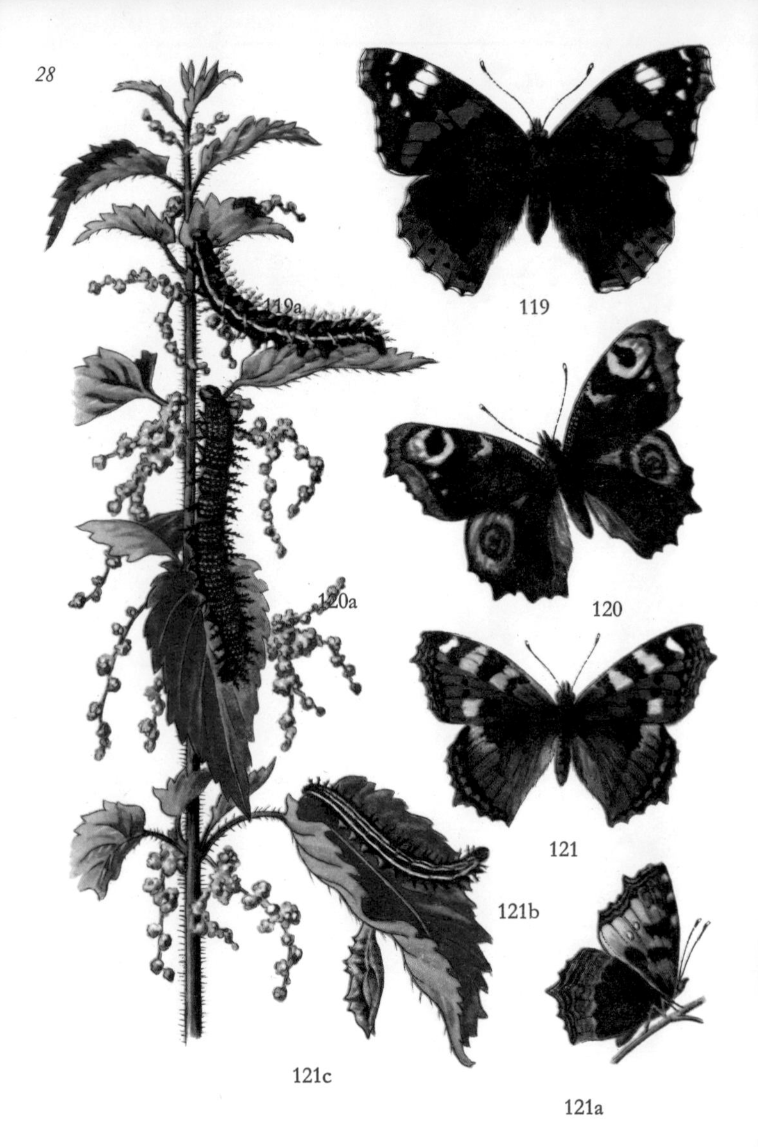

119 **Red Admiral,** *Vanessa atalanta* 119a caterpillar
120 **Peacock,** *Nymphalis io* 120a caterpillar
121 **Small Tortoiseshell,** *Aglais urticae* 121a under-side 121b caterpillar 121c chrysalis

The caterpillars of all three species feed on stinging nettles

122 **Camberwell Beauty,** *Nymphalis antiopa* 122a caterpillar 122b chrysalis 123 **Comma,** *Polygonia c-album* 123a caterpillar 124 **Queen of Spain Fritillary,** *Issoria lathonia* 124a under-side 124b caterpillar on wild pansy 125 **Dark Green Fritillary,** *Argynnis aglaia,* upper and under-side 126 **Silver-washed Fritillary,** *Argynnis paphia,* upper and undersides

127 **Scotch Argus,** *Erebia aethiops* 127a caterpillar 128 **Grayling** *Satyrus semele* 128a caterpillar 129 **Small Heath,** *Coenonympha pamphilus* 129a caterpillar 130 **Wall Brown,** *Pararge megaera* 130a caterpillar

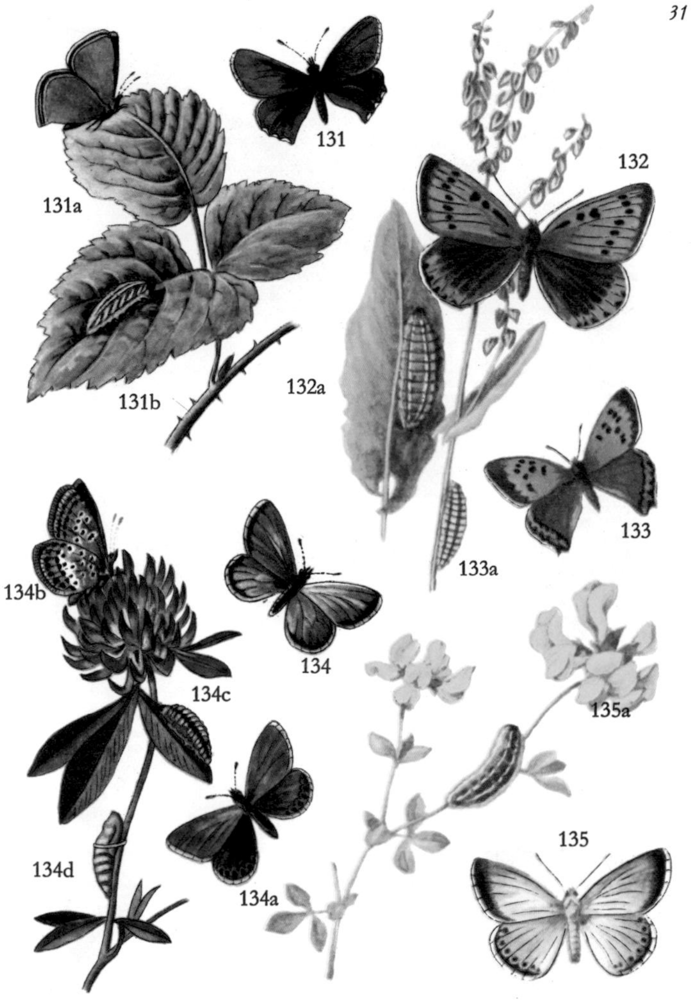

131 **Green Hairstreak,** *Callophrys rubi,* 131a under-side 131b caterpillar on bramble 132 **Large Copper,** *Lycaena dispar,* 132a caterpillar 133 **Small Copper,** *Lycaena phlaeas* 133a caterpillar 134 **Silver-studded Blue,** *Plebejus argus,* male 134a female 134b under-side 134c caterpillar 134d chrysalis 135 **Chalk Hill Blue,** *Lysandra coridon* 135a caterpillar

136 **Death's Head Hawk Moth,** *Acherontia atropos* 136a caterpillar
137 **Eyed Hawk Moth,** *Smerinthus ocellatus* 137a caterpillar
137b chrysalis

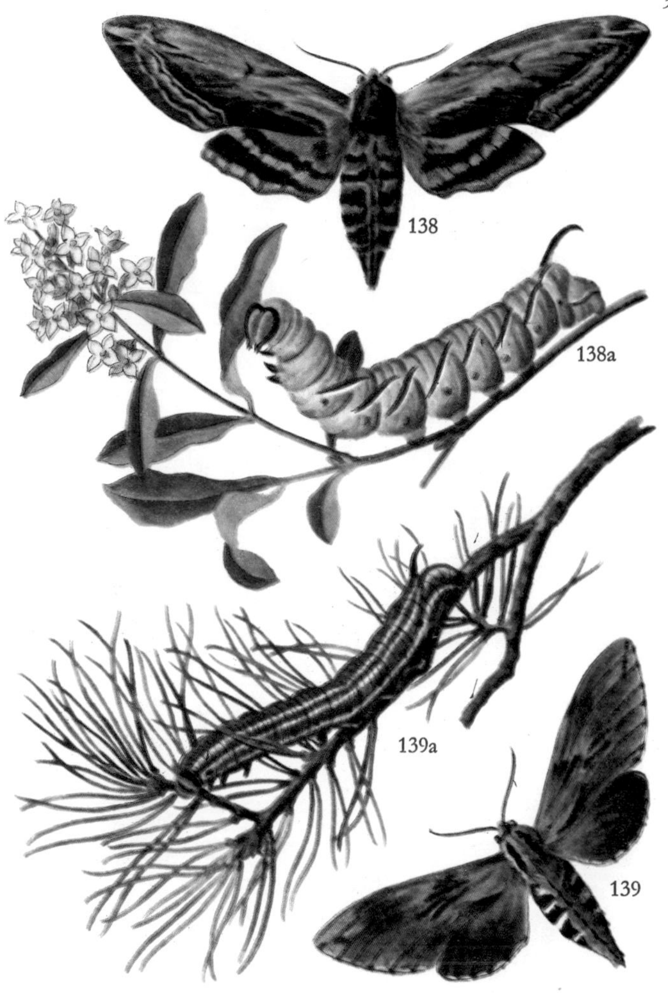

138 **Privet Hawk Moth,** *Sphinx ligustri* 138a caterpillar
139 **Pine Hawk Moth,** *Hyloicus pinastri* 139a caterpillar

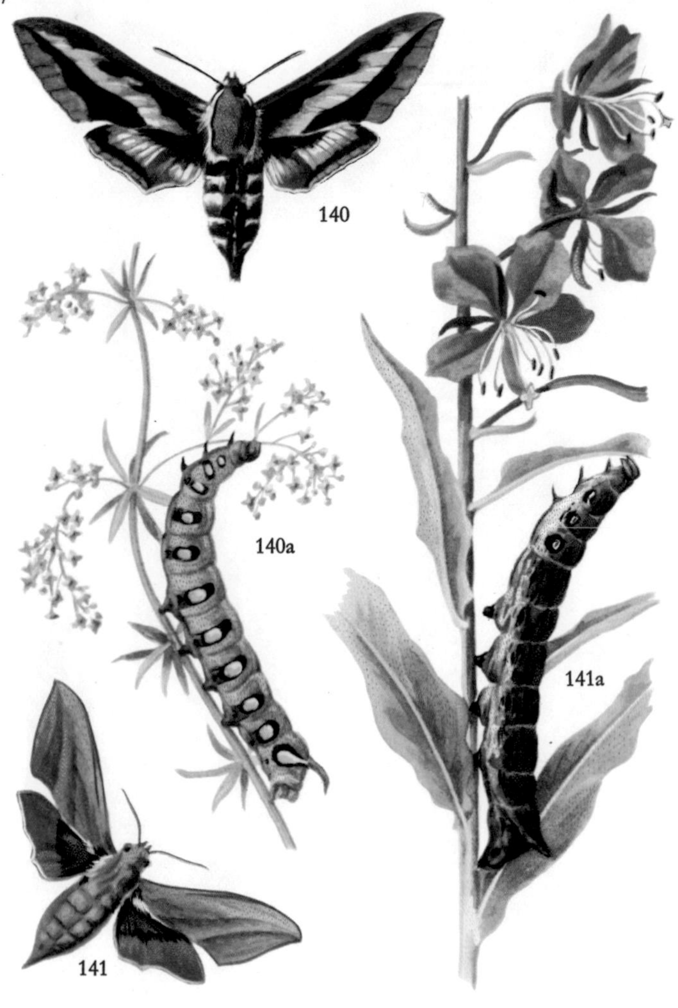

140 **Bedstraw Hawk Moth,** *Celerio galii* 140a caterpillar
141 **Elephant Hawk Moth,** *Deilephila elpenor* 141a caterpillar

142 **Puss Moth,** *Cerura vinula* 142a caterpillar 143 **Buff Tip Moth,** *Phalera bucephala* 143a caterpillar 143b chrysalis 144 **Vapourer Moth** *Orgyia antiqua,* male 144a female 144b caterpillar 144c chrysalis 145 **Small Eggar,** *Eriogaster lanestris,* male 145a female 145b egg bracelet on twig 145c caterpillar 145d cocoon

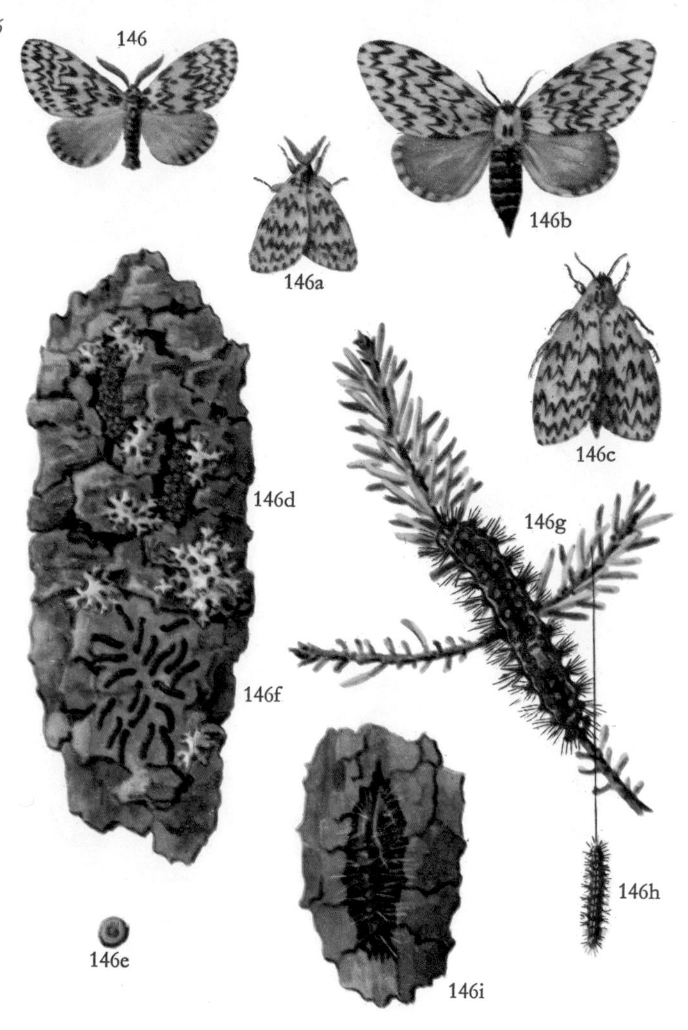

146 **Black Arches,** *Lymantria monacha,* male 146a male 146b and 146c female 146d egg batches on tree bark 146e egg enlarged 146f newly hatched caterpillars 146g full grown caterpillar 146h young caterpillar 146i chrysalis

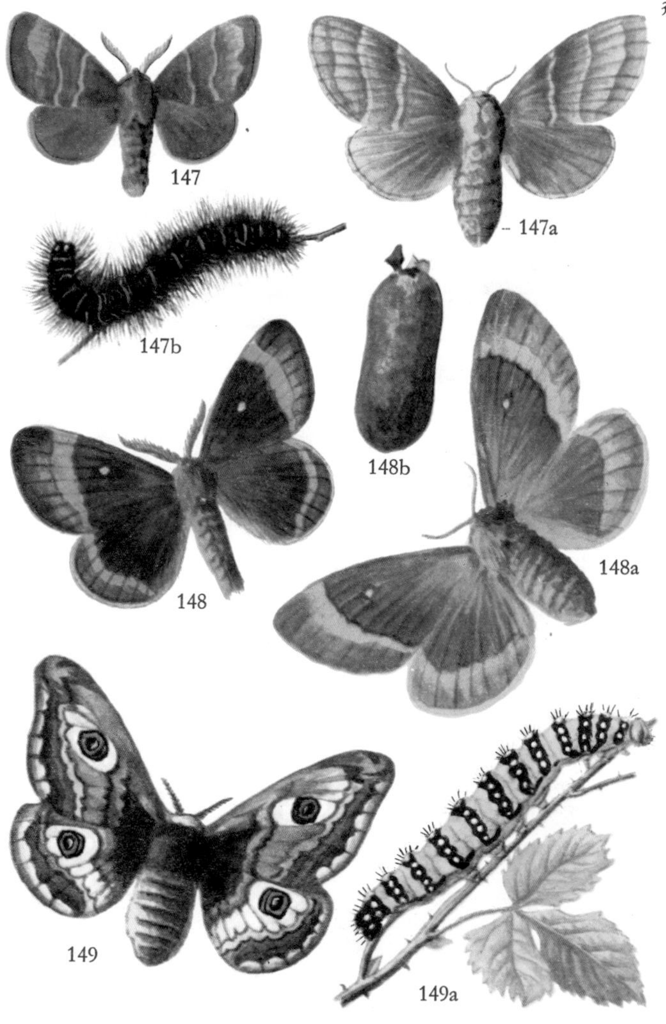

147 **Fox Moth,** *Macrothylacia rubi,* male 147a female 147b caterpillar
148 **Oak Eggar,** *Lasiocampa quercus,* male 148a female 148b cocoon
149 **Emperor Moth,** *Saturnia pavonia,* male 149a caterpillar

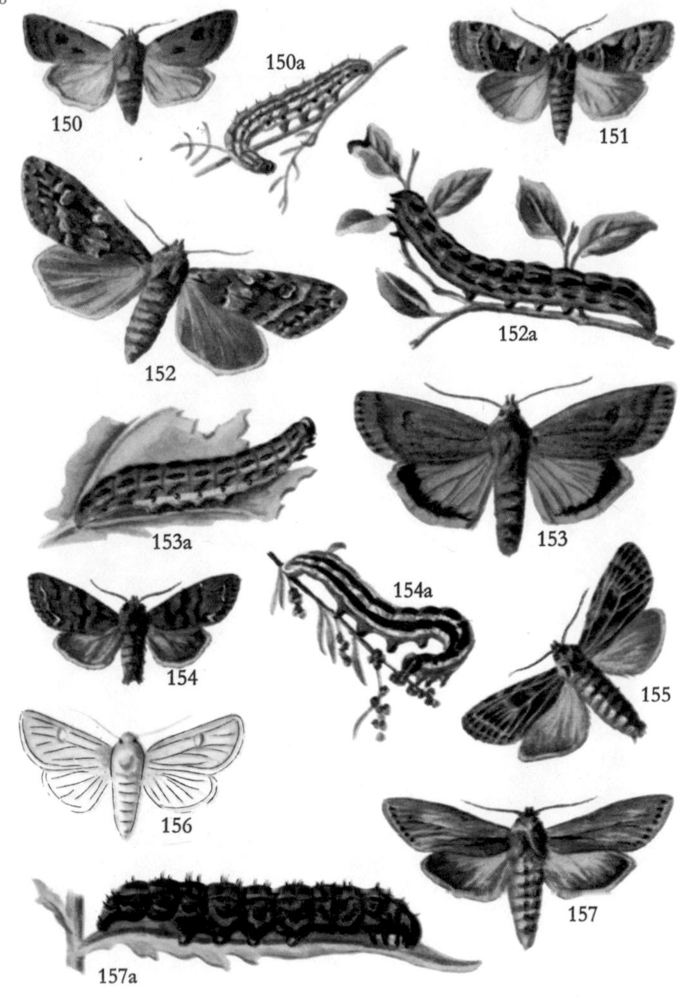

150 **Heart and Dart,** *Agrotis exclamationis* 150a caterpillar 151 **Setaceous Hebrew Character,** *Amathes c-nigrum* 152 **Great Brocade,** *Eurois occulta* 152a caterpillar 153 **Large Yellow Underwing,** *Triphaena pronuba* 153a caterpillar 154 **Broom Moth,** *Ceramica pisi* 154a caterpillar 155 **Feathered Gothic,** *Tholera popularis* 156 **The Claddagh,** *Luceria virens* 157 **The Shark,** *Cucullia umbratica* 157a caterpillar

158 **The Sallow,** *Cirrhia icteritia* 158a caterpillar 159 **Bird's Wing,** *Dypterygia scabriuscula* 160 **Clouded Border Brindle,** *Apamea crenata* 161 **Copper Underwing,** *Amphipyra pyramidea* 161a caterpillar 162 **The Dunbar,** *Cosmia trapezina* 162a caterpillar 163 **Silver Y.,** *Plusia gamma* 163a caterpillar 164 **Burnished Brass,** *Plusia chrysitis* 164a caterpillar

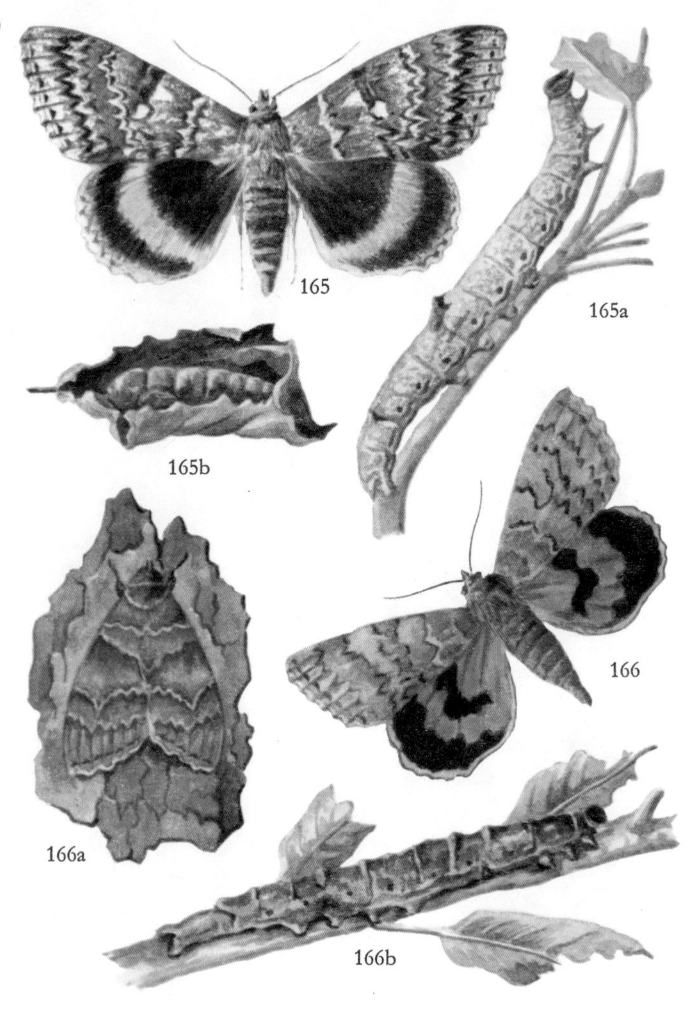

165 **Clifden Nonpareil,** *Catocala fraxini* 165a caterpillar 165b chrysalis
166 **Red Underwing,** *Catocala nupta* 166a moth at rest on tree trunk 166b caterpillar

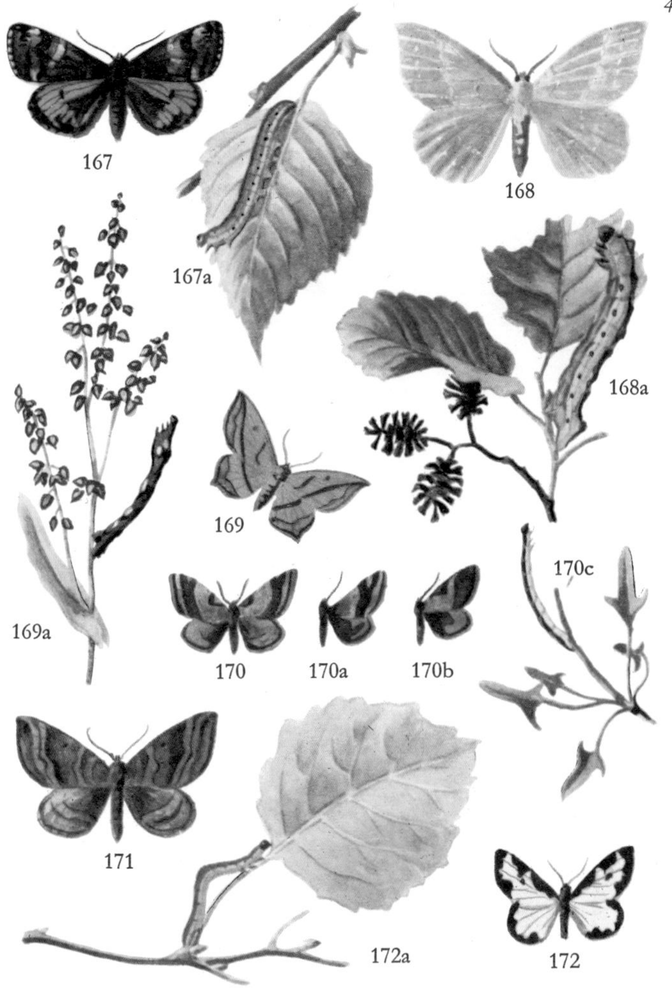

167 **Orange Underwing,** *Brephos parthenias* 167a caterpillar 168 **Large Emerald,** *Hipparchus papilionarius* 168a full grown caterpillar 169 **Blood Vein,** *Calothysanis amata,* 169a caterpillar 170 **Purple-barred Yellow,** *Lythria purpuraria* 170a and 170b colour variations 170c caterpillar 171 **Shaded Broad-bar,** *Ortholitha chenopodiata* 172 **Clouded Border,** *Lomaspilis marginata* 172a caterpillar

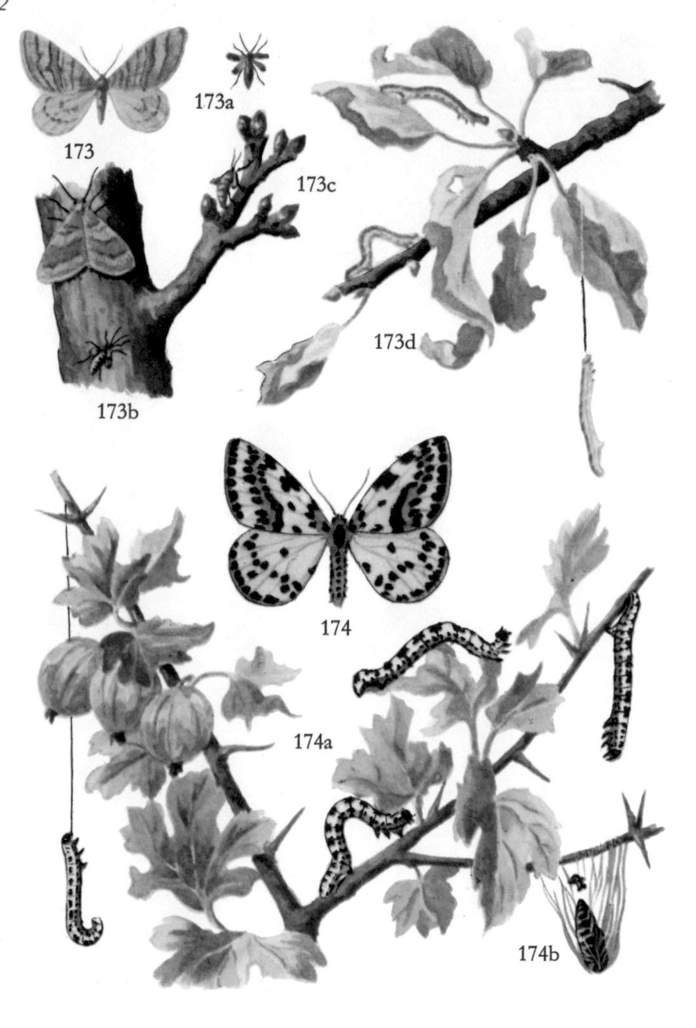

173 **Winter Moth,** *Operophtera brumata,* male 173a female 173b male and female at rest 173c female egg-laying 173d caterpillars
174 **Magpie or Currant Moth,** *Abraxas grossulariata* 174a caterpillars 174b chrysalis

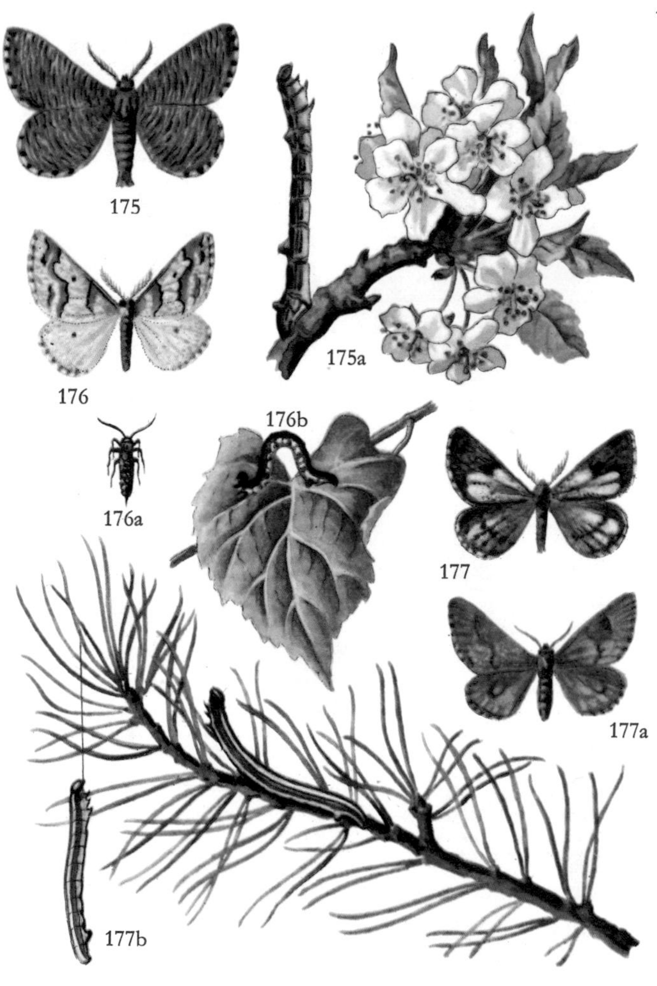

175 **Orange Moth,** *Angerona prunaria,* male 175a caterpillar
176 **Mottled Umber,** *Erannis defoliaria,* male 176a female
176b caterpillar
177 **Bordered White,** *Bupalus piniarius,* male 177a female
177b caterpillars

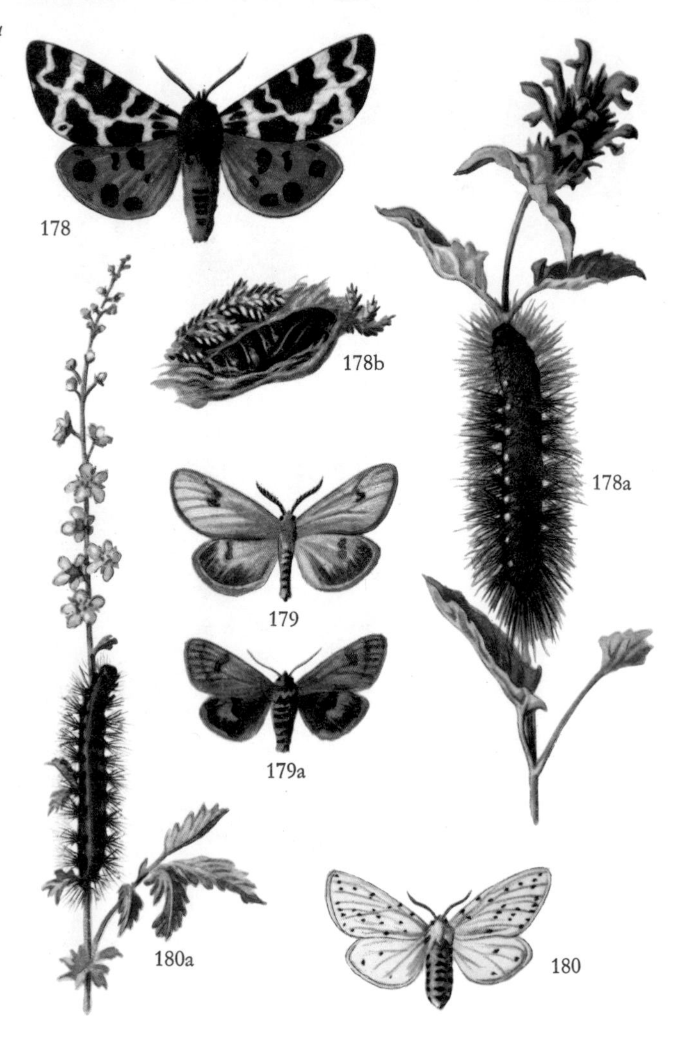

178 **Garden Tiger,** *Arctia caia* 178a caterpillar 178b chrysalis
179 **Clouded Buff,** *Diacrisia sannio,* male 179a female
180 **White Ermine,** *Spilosoma lubricipeda* 180a caterpillar

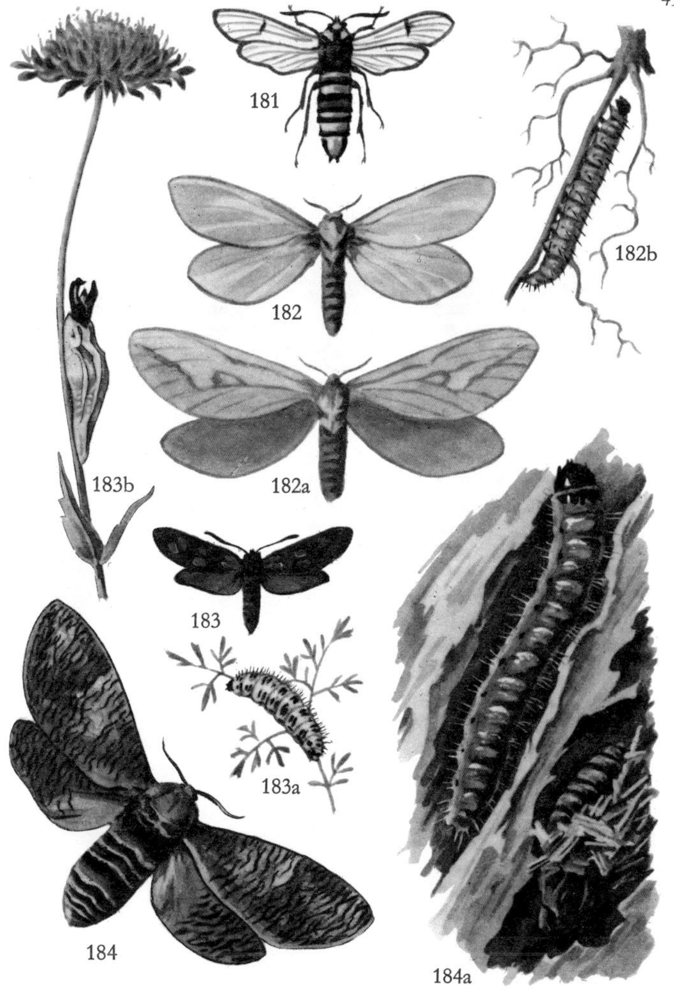

181 **Hornet Clearwing Moth,** *Sesia apiformis*
182 **Ghost Swift,** *Hepialus humuli,* male 182a female 182b caterpillar
183 **Six-spot Burnet,** *Zygaena filipendulae* 183a caterpillar 183b cocoon and empty chrysalis
184 **Goat Moth,** *Cossus cossus* 184a caterpillar and chrysalis

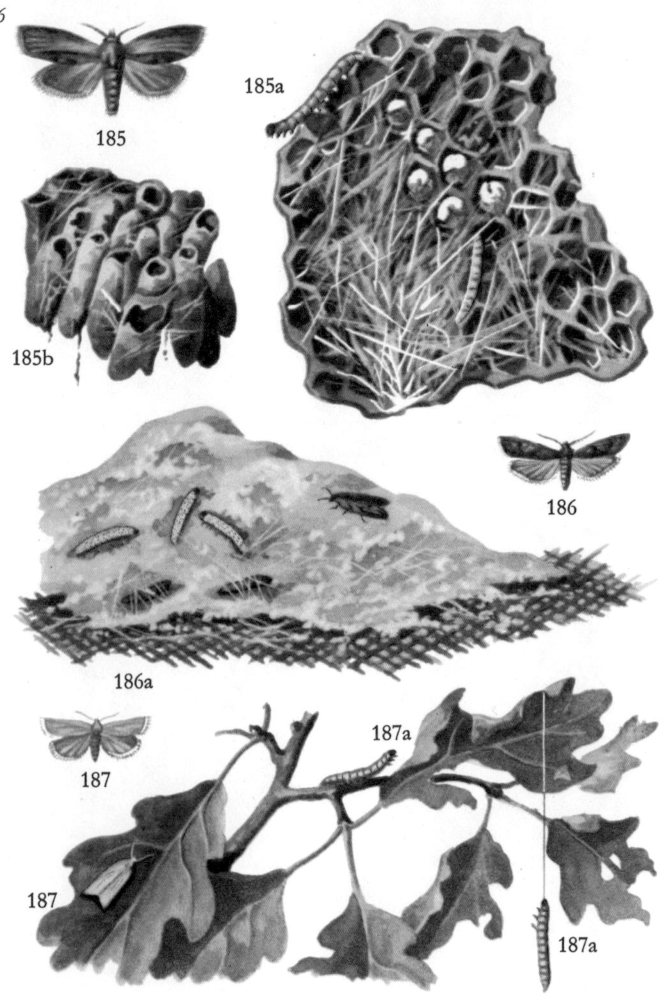

185 **Wax Moth,** *Galleria mellonella,* 185a caterpillar and web on honeycomb 185b cocoons amongst remains of comb

186 **Mill Moth** or **Mediterranean Flour Moth,** *Ephestia kuehniella* 186a caterpillar, chrysalis and moth on flour

187 **Green Oak Beauty,** *Tortrix viridana* 187a caterpillar

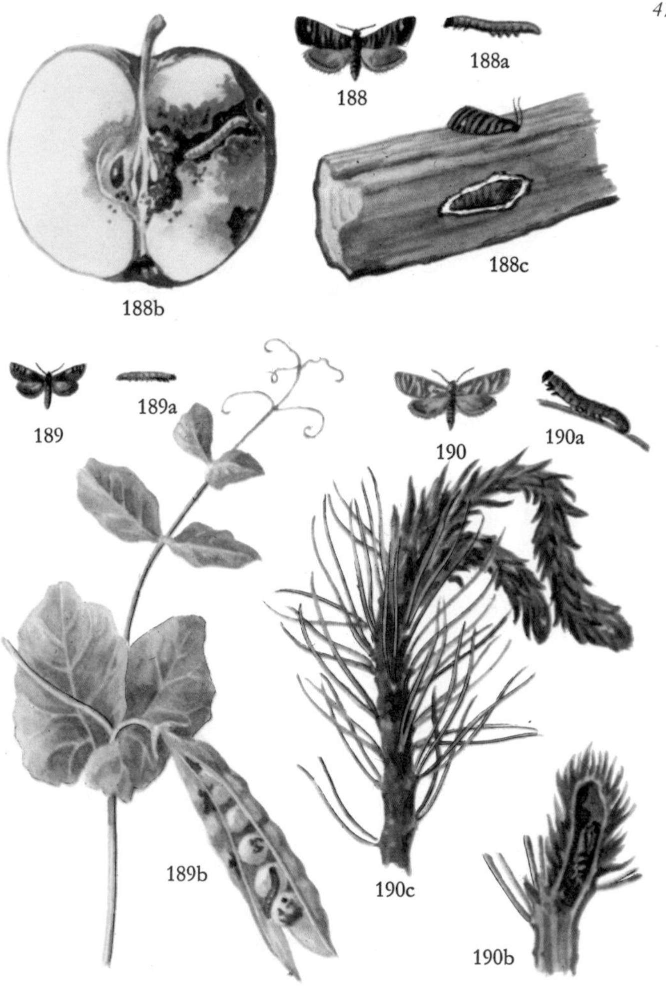

188 **Codling Moth,** *Cydia pomonella* 188a caterpillar 188b apple damaged by caterpillar 188c chrysalis 189 **Pea Tortrix,** *Laspeyresia nigricana* 189a caterpillar 189b caterpillars showing damage to pea pod 190 **Pine Tortrix,** *Evetria buoliana* 190a caterpillar 190b chrysalis in pine shoot 190c damaged and deformed pine shoot

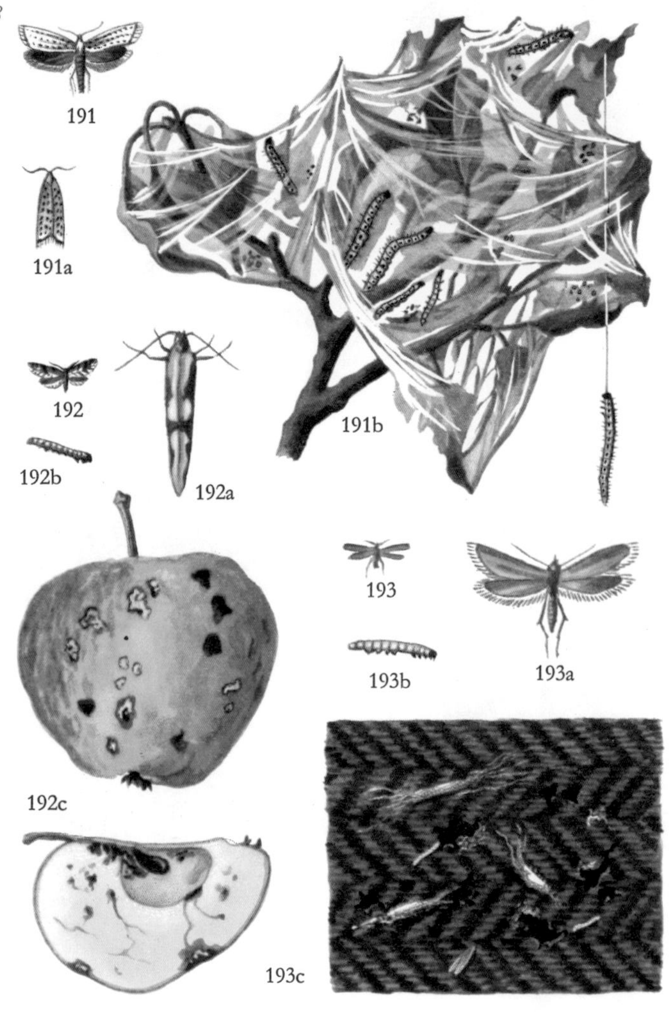

191 **Small Ermine,** *Hyponomeuta evonymella,* 191a at rest 191b web with caterpillars and cocoons 192 **Apple Fruit Miner,** *Argyresthia conjugella* 192a much enlarged 192b caterpillar 192c apples damaged by the caterpillars 193 **Clothes Moth,** *Tineola bisselliella* 193a moth much enlarged 193b caterpillar 193c caterpillars and cocoons on damaged cloth

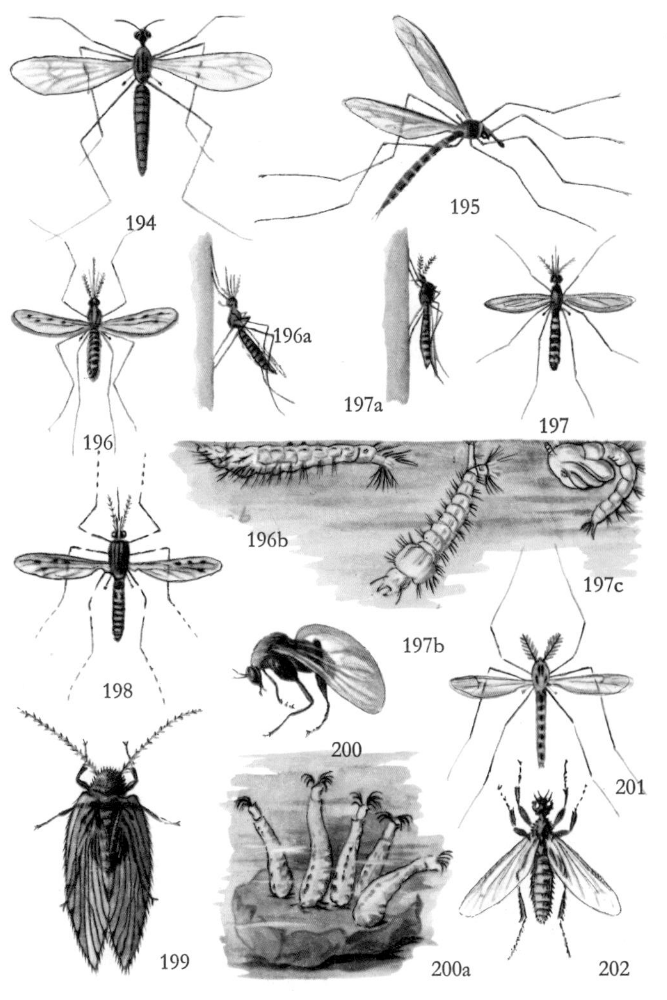

194 **Winter Gnat,** *Trichocera hiemalis* 195 **Common Daddy Long Legs,** *Tipula oleracea* 196 **Spangle-winged Mosquito,** *Anopheles maculipennis* 196a in resting position 196b larva 197 **Common Gnat,** *Culex pipiens* 197a in resting position 197b larva 197c pupa 198 **Ringed Mosquito,** *Theobalida annulata* 199 **Hairy Moth Fly,** *Psychoda alternata* 200 **River Fly or Buffalo Gnat,** *Simulium reptans* 200a larvae 201 **Harlequin Fly,** *Chironomus plumosus* 202 **Red-legged Bibio,** *Bibio pomonae*

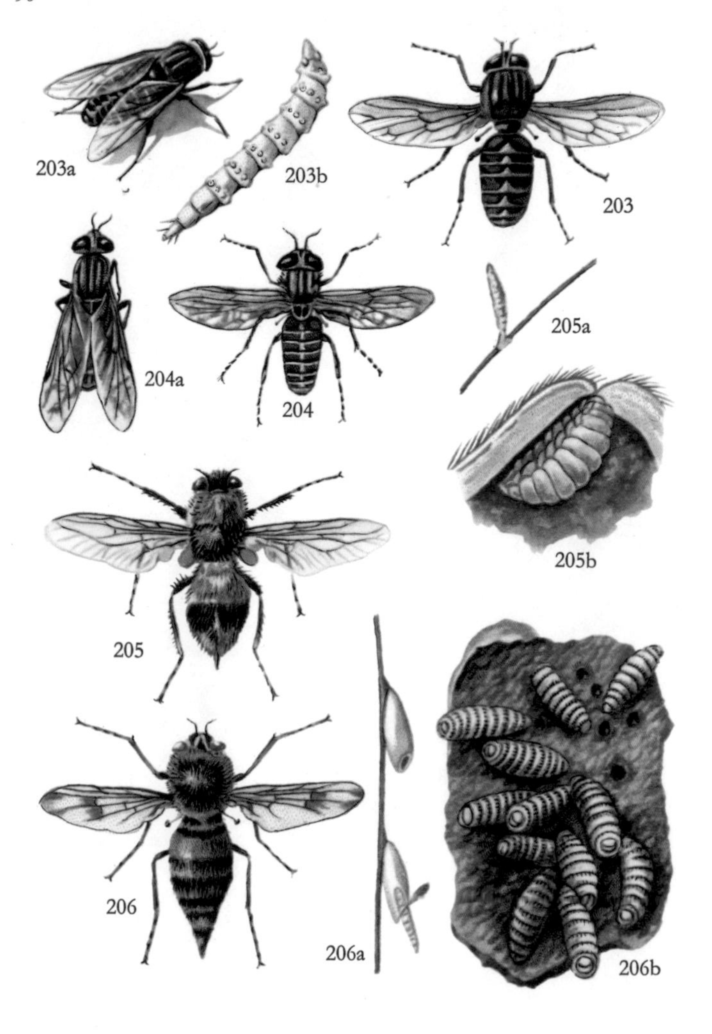

203 **Horse Fly or Gad Fly,** *Tabanus bovinus* 203a in resting position 203b larva 204 **Clegg,** *Haematopota pluvialis* 204a in resting attitude 205 **Warble Fly,** *Hypoderma bovis* 205a egg on hair of host 205b larva under skin of host 206 **Common Horse Bot Fly,** *Gasterophilus intestinalis* 206a eggs on hair of host 206b part of stomach-wall of host with maggots in position

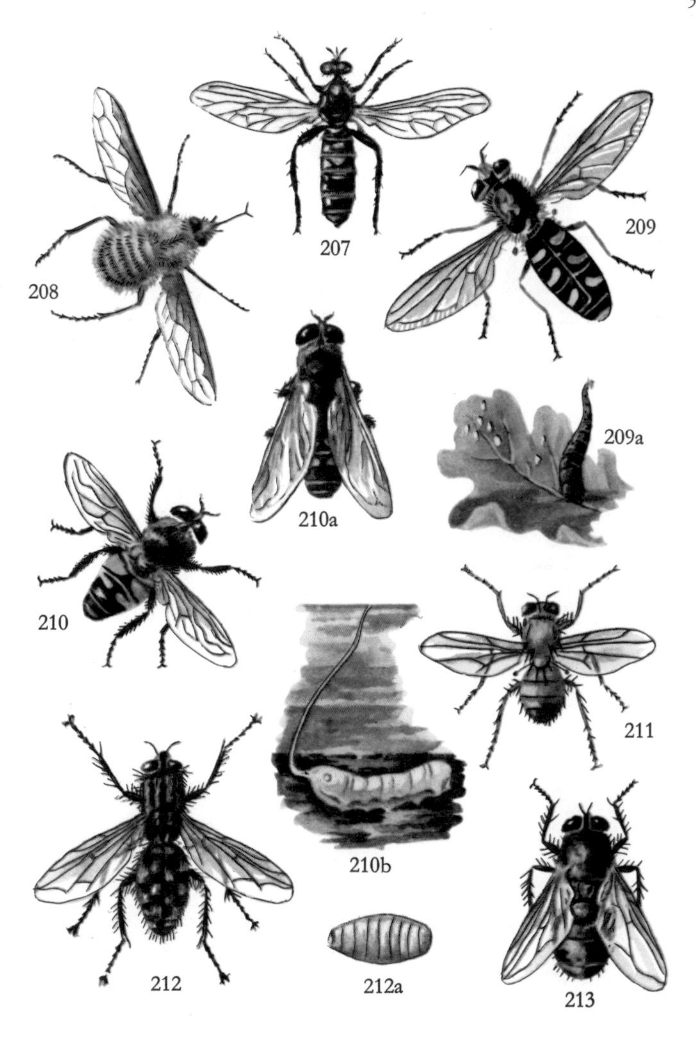

207 **Assassin Fly,** *Laphria marginata* 208 **Large Bee Fly,** *Bombylius major* 209 **Swarming Hover Fly,** *Scaeva pyrastri* 209a larva feeding on aphids 210 **Drone Fly,** *Eristalis tenax* 210a in resting attitude 210b larva 211 **Vinegar Fly,** *Drosophila transversa,* much enlarged 212 **Flesh Fly,** *Sarcophaga carnaria* 212a pupa 213 **Green Bottle Fly,** *Lucilia caesar*

214 **Blow Fly,** *Calliphora vomitaria* 215 **Deer Bot Fly,** *Cephenomyia auribarbis* 215a larva 216 **House Fly,** *Musca domestica* 216a in cleaning attitude 216b maggot 216c puparium 217 **Lesser House Fly,** *Fannia canicularis* 218 **Forest Fly,** *Hippobsca equina* .218a in resting attitude 219 **Sheep Ked,** *Melophagus ovinus* 220 **Dog Flea,** *Ctenocephalides canis* 221 **Flea,** *Pulex irritans* 221a larva 221b pupa

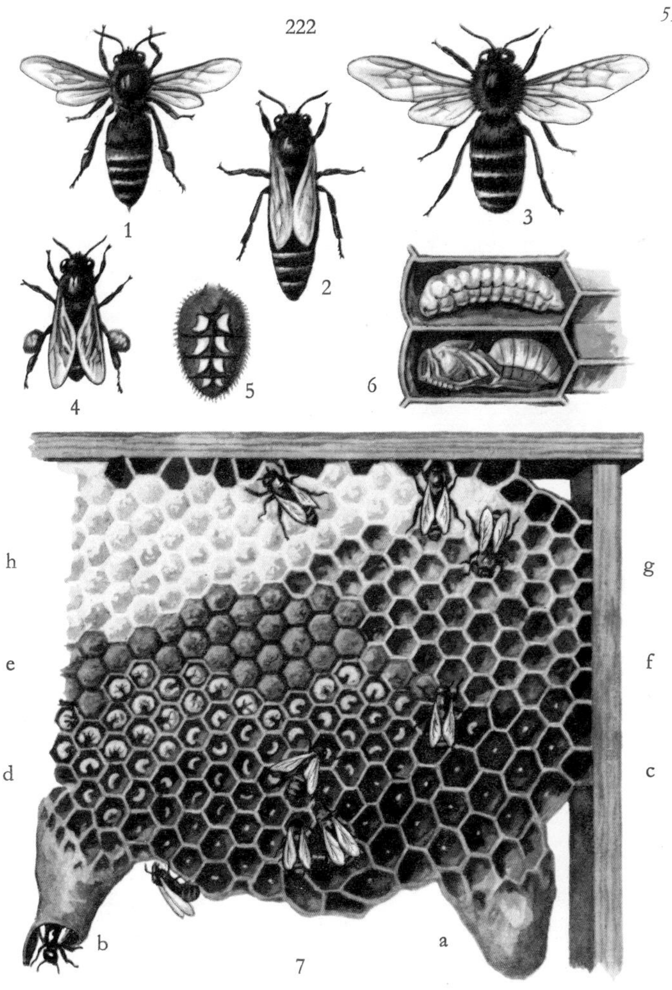

222 **Honey Bee or Hive Bee,** *Apis mellifica* 1 Worker 2 Queen 3 Drone 4 Worker with laden pollen baskets 5 Abdomen of worker, showing wax scales on under-side 6 Larva and pupa in cells 7 Comb with a) closed b) open queen cell with newly hatched queen emerging c) drone cells d) worker cells with eggs and larvae in various stages e) sealed brood cells f) pollen cells g) open and h) sealed honey cells

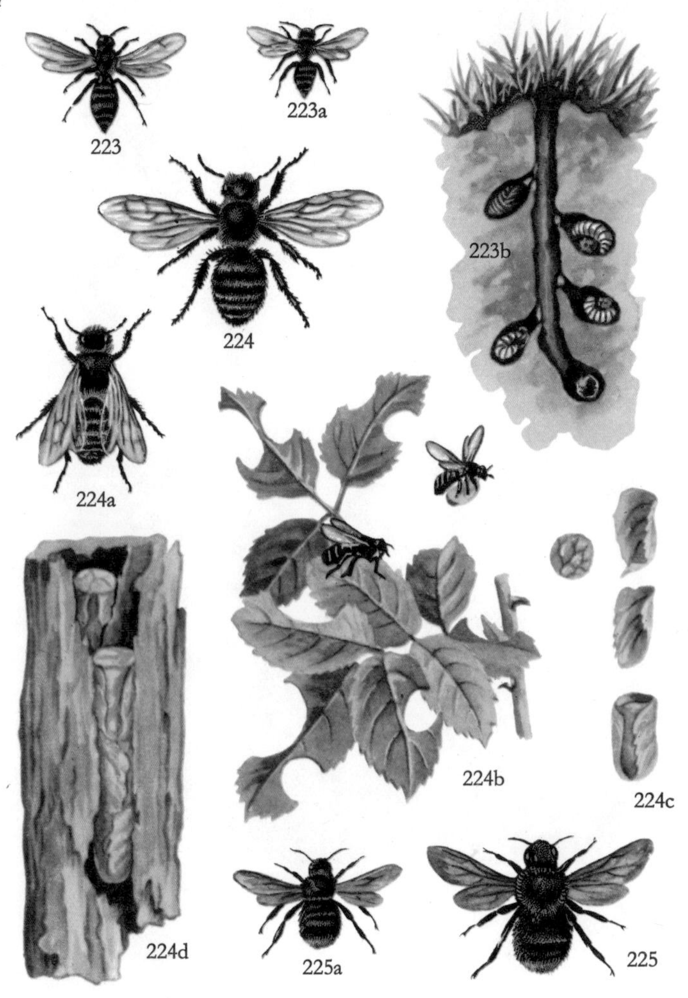

223 **Early Mining Bee,** *Andrena albicans,* male 223a female 223b shaft in soil with larval chambers 224 **Patchwork Leaf-cutter Bee,** *Megachile centuncularis,* female 224a male 224b rose leaves, showing cuts made by bees 224c portions of leaves cut to form cells 224d cells in position in hollow wood 225 **Hill Cuckoo Bee,** *Psithyrus rupestris,* female 225a male

226 **Buff-tailed Bumble Bee,** *Bombus terrestris* 1 Female 2 Worker 3 Male 4 Underground nest

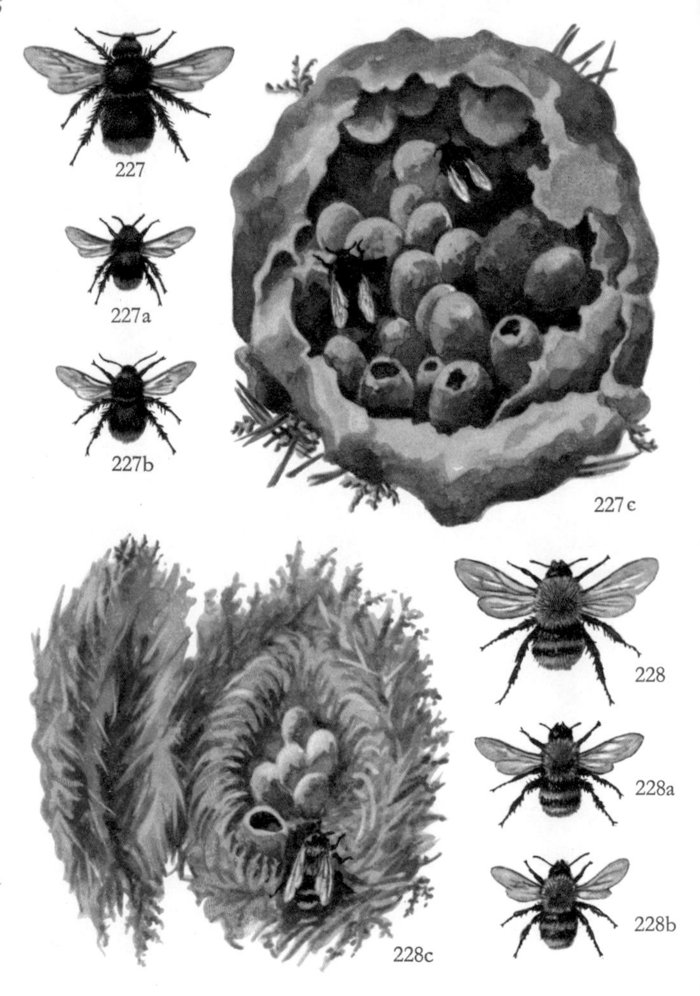

227 **Red-tailed Bumble Bee,** *Bombus lapidarius,* female 227a worker 227b male 227c nest partly opened to show interior 228 **Common Carder Bee,** *Bombus agrorum,* female 228a worker 228b male 228c nest made of grass and moss, opened to show queen and first brood cells

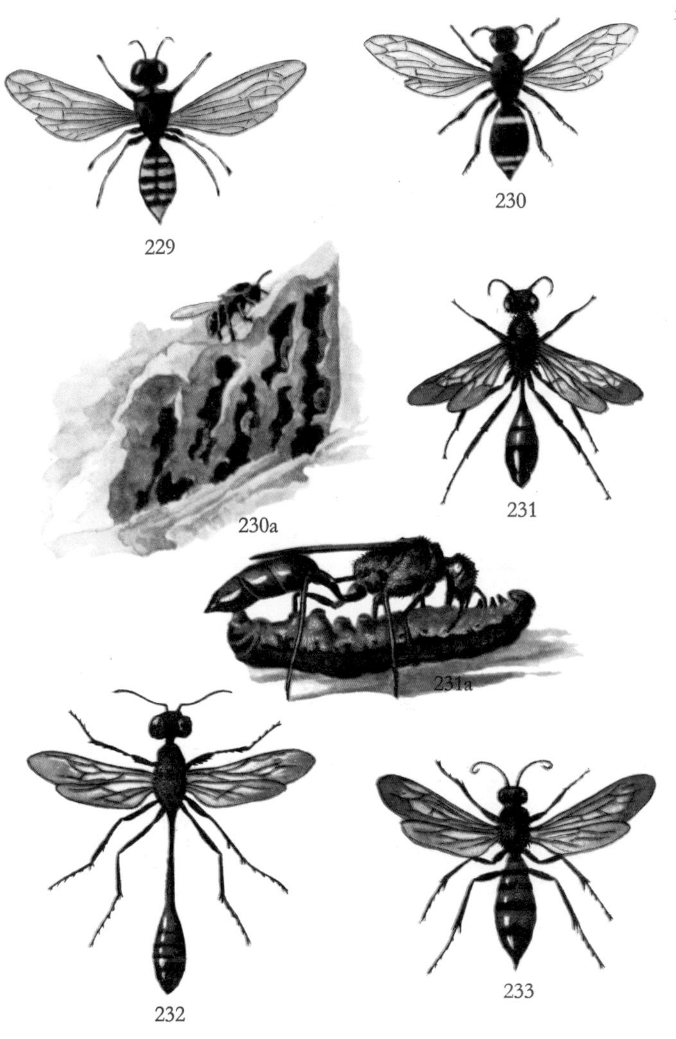

229 **Big-headed Digger Wasp,** *Ectemnius cavifrons* 230 **Wall Mason Wasp,** *Odynerus parietum* 230a nest 231 **Hairy Sand Wasp,** *Podalomia viatica* 231a stinging caterpillar of night-flying moth 232 **Red-banded Sand Wasp,** *Sphex sabulosa* 233 **Red-banded Spider Wasp,** *Anoplius fuscus*

234 **Common Wasp,** *Vespa vulgaris* 1 Female 2 Worker 3 Male
4 partly opened underground nest

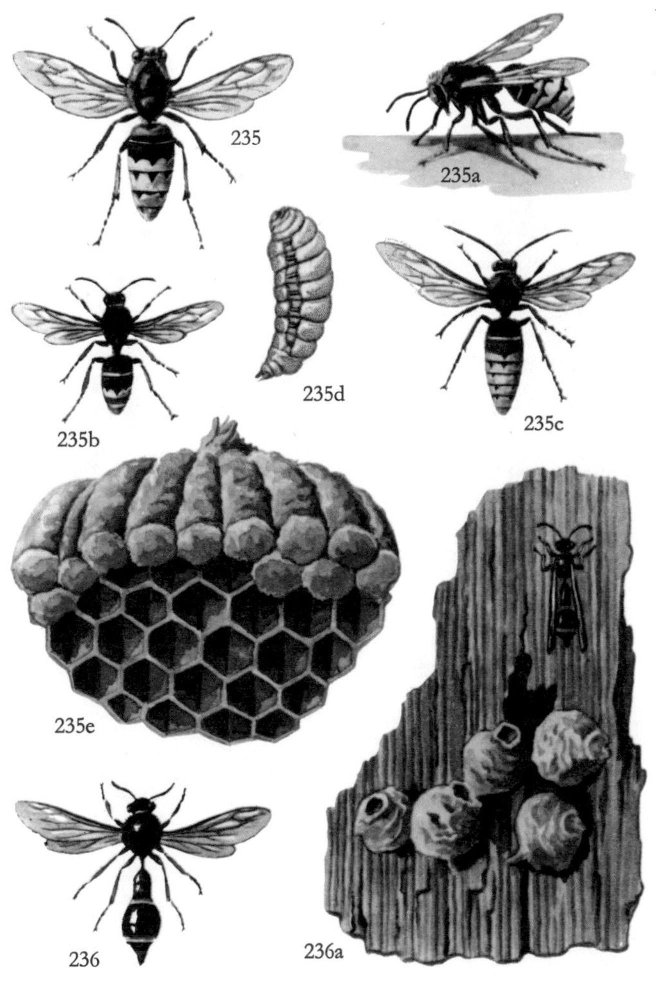

235 **Hornet,** *Vespa crabro,* female 235a in profile 235b worker 235c male 235d larva 235e part of the nest with open and closed cells
236 **Heath Potter Wasp,** *Eumenes coarctata* 236a the wasp in resting attitude, with three open and two closed cells

237 **Wood Ant,** *Formica rufa* 1 Winged female 2 and 3 Workers 4 Winged male 5 Section of underground portion of nest showing a) eggs, b) larvae, c) pupae, 6 anthill above the nest

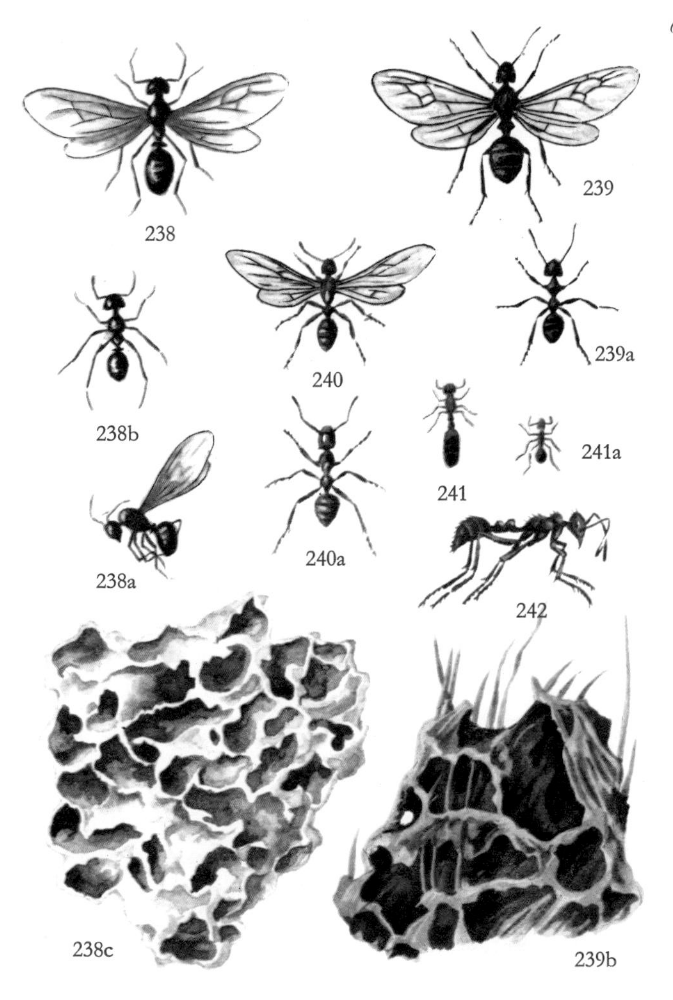

238 **Jet Ant,** *Lasius fuliginosus,* female 238a female profile 238b worker 238c part of nest 239 **Small Black Ant,** *Lasius niger,* female 239a worker 239b part of nest, built among grass 240 **Yellow Ant,** *Lasius flavus,* male 240a female 241 **Pharaoh's Ant** *Monomorium pharaonis,* female, 241a worker 242 **Red Ant,** *Myrmica rubra,* worker

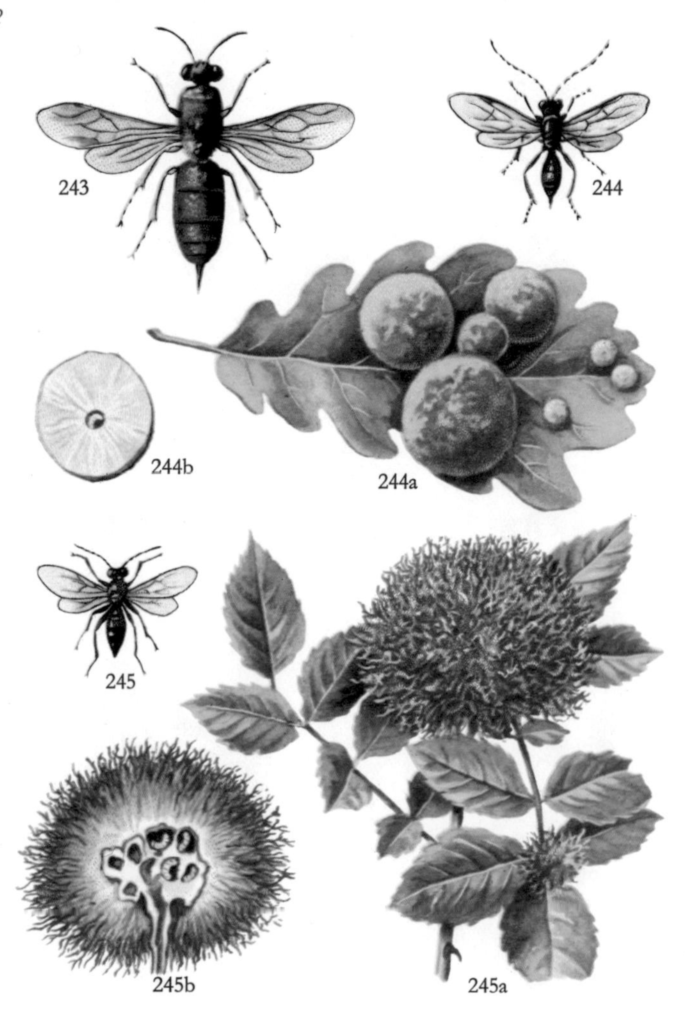

243 **Ruby-tailed Wasp,** *Chrysis ignita*
244 **Cherry Gall,** *Cynips quercusfolii* 244a cherry galls on oak leaf
224b cut gall to show larva
245 **Robin's Pin Cushion,** *Rhodites rosae* 245a the gall on rose stem
245b gall cut open to show larvae in chambers

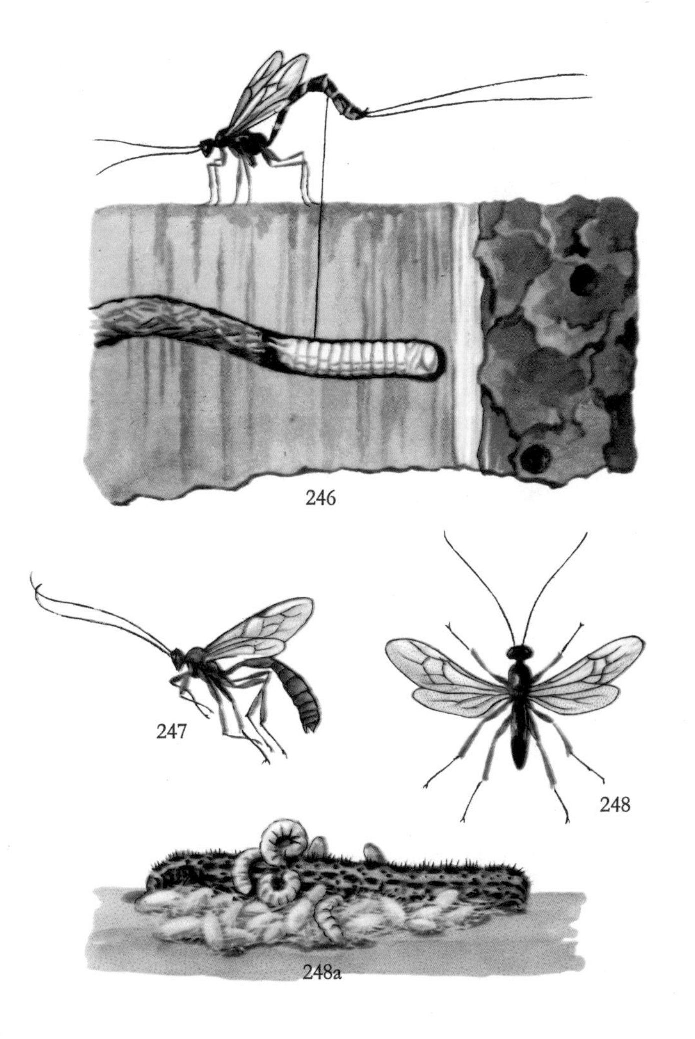

246 **Horn-tail Ichneumon,** *Rhyssa persuasoria,* female laying egg on larva of Giant Wood Wasp

247 **Yellow Ophion,** *Ophion luteus*

248 **Cabbage White Wasp,** *Apanteles glomeratus* 248a grubs emerging from Cabbage White caterpillar, and their cocoons

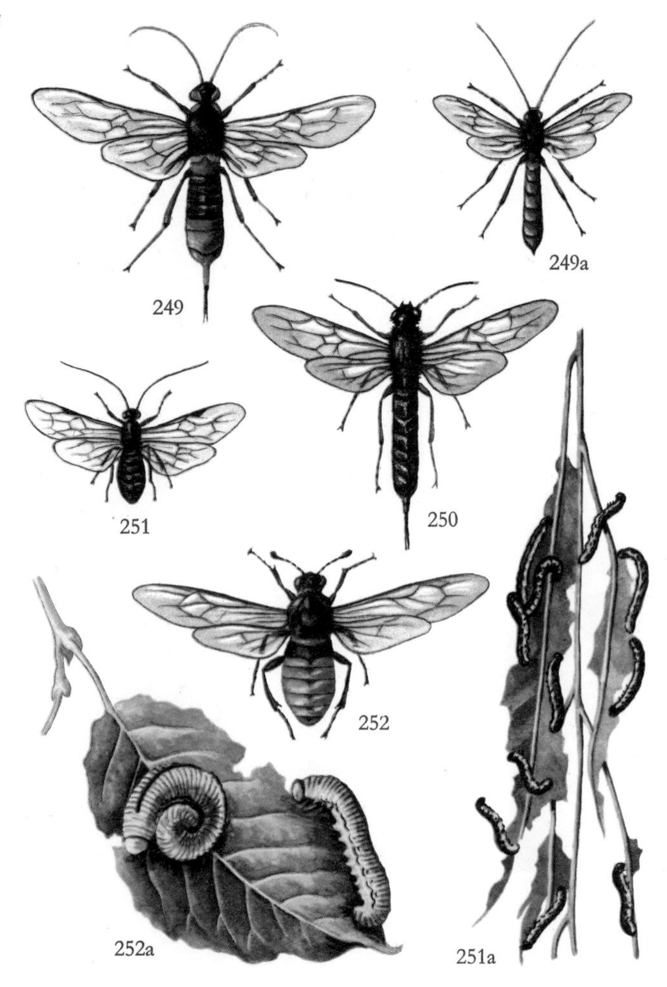

249 **Giant Wood Wasp,** *Sirex gigas,* female 249a male
250 **Steel-blue Wood Wasp,** *Sirex juvencus*
251 **Willow Sawfly,** *Pteronidea salicis* 251a larvae on willow leaves
252 **Yellow Cimbex,** *Cimbex lutea* 252a larvae on sallow

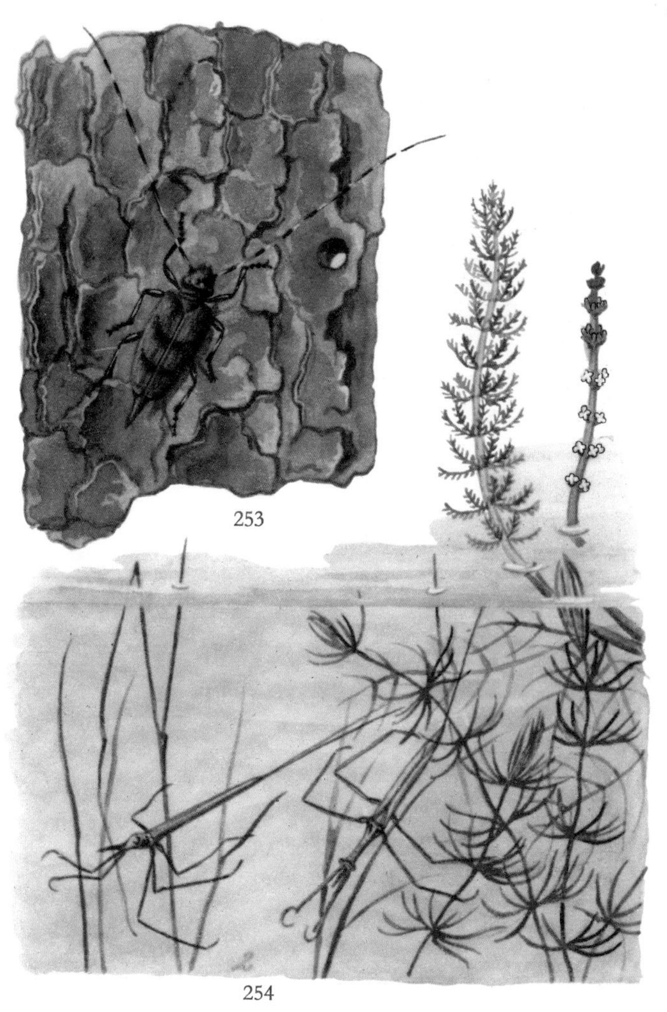

PROTECTIVE RESEMBLANCE

253 **Timberman,** *Acanthocinus aedilis* on bark
254 **Narrow Water Scorpion,** *Ranatra linearis,* among pond weeds

PROTECTIVE RESEMBLANCE

255 **Grayling,** *Satyrus semele* on ground amongst thyme 256 **Peppered Moth,** *Biston betularia* on birch bark 257 **Stick caterpillar of Phoenix Moth,** *Lygris prunata* on currant stem 258 **Hornet Clearwing Moth,** *Sesia apiformis* and 259 **Swarming Hover Fly,** *Myiatropa florea* mimicking 260 **The Common Wasp,** *Vespa vulgaris,* male 260a female 261 **Wasp** (on right) and **Hover Fly** (on left) on flowerhead

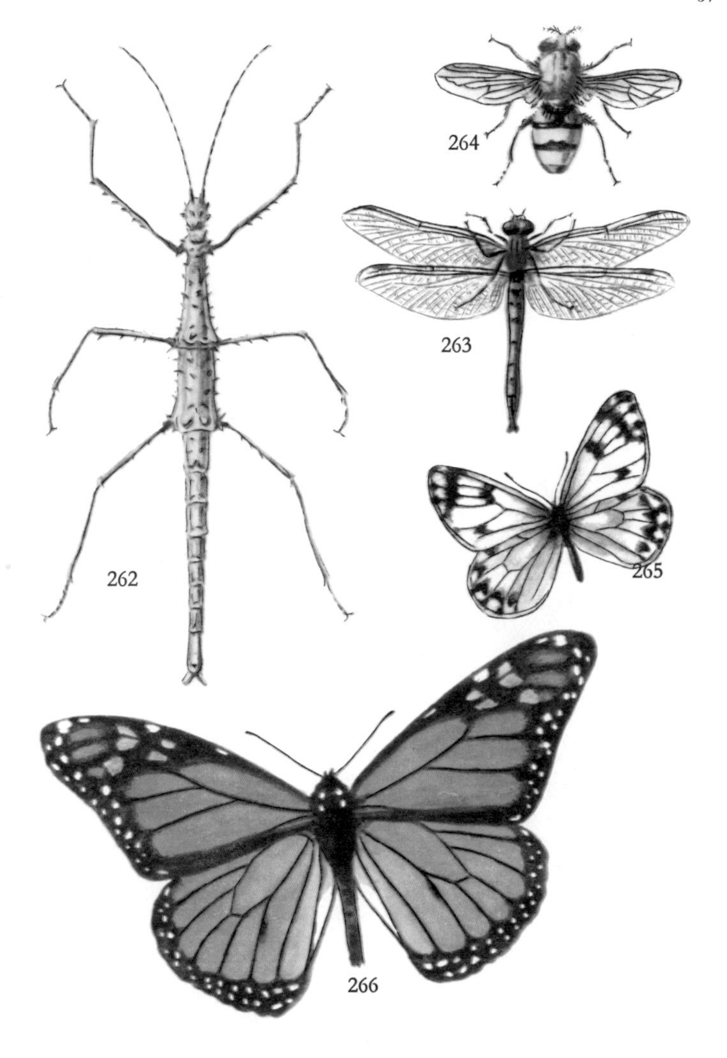

RARE VISITORS

262 **New Zealand Stick Insect,** *Acanthoxyla prasina* 263 **Red-veined Dragonfly,** *Sympetrum fonscolombii* 264 **Banded Hover Fly,** *Volucella zonaria* 265 **Bath White,** *Pontia daplidice* 266 **Milkweed Butterfly,** *Danaus plexippus*

267 **Colorado Beetle,** *Leptinotarsa decemlineata* 1 beetle 2 larva 3 pupa 4 egg-mass (all enlarged) 5 beetle and pupa on potato haulms (natural size)

INTRODUCTION

In the popular mind, almost any small creeping, crawling or flying creature is an insect. To naturalists, however, the term is a precise one having a definite application. The celebrated Swedish scientist, Carl von Linné, who in 1758 published his *Systema Naturae*, and thus provided naturalists with the method of naming and describing animals that is still in use, applied the term *Insecta* to the crabs, shrimps and lobsters, spiders and mites, centipedes and millipedes as well as the true insects, because all these have bodies made up of rings or segments.

True insects, when adult, can be distinguished from these others because they have the following combination of characters:- (1) a separate head bearing one pair of antennae (feelers), eyes, and three pairs of mouth parts, (2) a thorax composed of three segments more or less fused together each bearing one pair of legs and, in the case of the second and third segments, also as a rule a pair of wings, (3) a body composed of about ten visible segments and devoid of legs. These features can be seen in the outline figures of a ground beetle given on page 73 provided it is remembered that the beetle's elytra (5 in the right hand figure) correspond with the forewings of a moth or butterfly, and that the hind wings are hidden beneath them. The three sets of mouth parts are indicated by the letters A, B and B′ in the left hand figure, A being the jaws and B and B′ the two sets of palpi used as organs of touch.

In a class of animals so numerous both in individuals and in species as the insects, an almost endless range of variation in the shape and function of these principal structures has developed during their immense period of evolution. Some insects have lost the front pair of wings, many the hind wings, and even more have none at all. In many butterflies, such as the tortoiseshells, the front legs are reduced to small brush-like organs; some parasitic insects, even when adult, are devoid of legs. The comparatively simple mouth parts of such an insect as the ground beetle illustrated, or a cockroach, have been modified to form the piercing and sucking organs of bugs and mosquitoes and the long coiled proboscis of a butterfly or moth. Several different types of antennae are illustrated on page 73.

Insects differ from nearly all other animals in that their hard parts are on the outside. They have no internal skeleton, only an external one, called an exoskeleton. This is composed mainly of a horny material called chitin, very tough and durable. To the inner surfaces of this are attached all the muscles that are used in movement, in feeding, egg-laying etc. Along the sides of this 'case' as it might be called there is to be seen a series of very small openings, called spiracles, one on either side of most segments, through which the insect breathes. It has no lungs, the air being distributed throughout the body by means of minute branching tubes. Along the lower surface of the body there runs a double nerve cord, similarly branching in each segment, and broadening and thickening in the head and uniting above the mouth to form a brain. The alimentary canal (gut) through which food passes during digestion, in its simplest form, is little more than a tube running from the mouth to the tip of the abdomen. All the internal parts are bathed in a blood-fluid which is usually colourless and is circulated by means of a heart which takes the form of a tube lying along the insect's back; its pulsations can easily be seen through the semi-transparent skin of many caterpillars.

Like most animals, insects begin life as eggs. These are, in the great majority of cases, laid by the females on or near the food that the young larvae (grubs, maggots, caterpillars etc.) will need on hatching. However, in insects there are many exceptions; reference is made in the descriptions of the figures, for example, to the fact that in many flies and aphids, the eggs develop within the body of the female into larvae which are born alive. Fertilisation of the eggs is normally secured only by the union of the sexes; here again some female insects depart from the normal and reproduce their kind without any intervention of the male, a phenomenon known as parthenogenesis or virgin birth, and very common among aphids and sawflies.
The grub that hatches from the egg may look very much like the full-grown insect that laid the egg (a young grasshopper for example) or it may, if it is a fly maggot or a caterpillar, look utterly different. In the most primitive kinds of insect, such as *Lepisma* (Fig. 1), which never have wings, the young are indistinguishable from the adults, except in size, and there are no visible external changes during growth; such insects are said to be ametabolous. Another type of development is well illustrated by the grasshoppers (Fig. 7) and bugs (Figs. 16/20). In these at each moult the young larva comes increasingly to resemble the adult, especially in the external growth of the wings. Such larvae are generally called nymphs and the type of development they show is said to be hemimetabolous, which implies that there are visible changes of form during growth but not such complete differences between the stages as in the next group. This third and last group contains the bulk of the insect world, since it includes the beetles, butterflies and moths, flies, ants, beetles and wasps. In these development is complete (holometabolous) since the four stages through which the insect passes, egg → caterpillar (or maggot) → chrysalis (or pupa) → adult are utterly different and distinct.
Whatever course of development an insect pursues from egg to adult, the sole functions of the larva are eating and growing. Once the adult stage has been reached, no further growth occurs: small flies do not, for example, grow into larger flies, nor small moths, like clothes moths, into bigger moths. The difficulty of growth whilst enclosed in a tough, horny exoskeleton of chitin, which does not expand, is overcome by the process of moulting. For example, when the skin of a caterpillar has become too tight for it, it moults, like a snake. At the moment of moulting the underlying chitin is soft and readily expands to the new size required; but it quickly hardens on exposure to the air. The number of moults that occur during growth varies enormously. Most caterpillars moult five or six times; some beetle grubs, especially when in a not very favourable situation for feeding, may moult twenty or more times. Insects with the incomplete type of metamorphosis (change of form during growth) are generally active throughout life, the last larval moult leading directly to the adult. A very important feature of those with complete metamorphosis is the addition of the pupal stage. This is interposed between the last larval stage and the adult and is a resting stage during which the internal tissues are broken down and re-formed in the shape of the butterfly, bee, beetle or fly which will eventually emerge from the pupa or chrysalis.
To fit them for life in their special environments insects have the senses of sight, smell, taste, touch and hearing. Possibly they have other senses, for they are richly endowed with sense organs, most of which are exceedingly minute. Both larvae and adults have, as a general rule, simple eyes of a kind that can do little more than appreciate light intensities. The compound eyes, which are so conspicuous a

feature of moths, dragonflies and many other insects, consist of numbers (sometimes running to thousands) of separate optical units, each one represented by one of the facets on the surface of the eye. The mosaic that results from the combination of this large number of sectional pictures appears to be adequate: one has only to watch a dragonfly in flight to be convinced. That at least some butterflies and bees can distinguish some colours has been proved; and it is known also that in many insects vision extends much further into the ultra-violet range of the spectrum than in man, a fact made use of by collectors who use ultra-violet light as a means of attracting night-flying insects.

There is no doubt, however, that the senses of smell, taste and touch are those most important in an insect's life, provided it escapes its enemies by seeing or hearing them in time. The organs of touch are principally minute sensory hairs scattered widely over the surface of the body and limbs. The organs that perceive scents (sensillae) are so small that 30,000 of them may be present on a single antenna of a bee; they occur chiefly on the antennae and are highly sensitive, though probably limited in scope. It is certain that most moths and probably most other insects, find their mates by the sense of smell and the same sense is very important in the search for food. The well-known collector's trick of 'assembling' males of Eggar moths (and others) depends upon the ability of the male to detect the scent of the female at a range said often to exceed a mile. In hive bees too, and amongst ants, it is of the utmost importance; indeed the whole life of the colony in the case of ants depends on it. The organs of taste are even more widely distributed than those of smell; though mainly found in the region of the mouth, on palpi and proboscis for example, they are also often present on the feet.

More obscure are the senses, believed to rest mainly in certain delicate static organs in the antennae, by means of which an insect maintains its balance and position when in flight, when swimming or otherwise moving about. If the antennae of a butterfly are removed its flight control is completely destroyed; similarly if those of a Whirligig Beetle are removed it can no longer avoid colliding with objects on the surface of the water on which it swims. As very many insects have a capacity for producing sounds (stridulation), which are not always perceptible to the human ear, it is a legitimate assumption that they should also have the means of hearing them. It is by means of the latter that ants and some caterpillars perceive the sounds that cause them sometimes to 'stop in their tracks' and 'freeze'.

Reference was made above to insects taking to flight to escape their enemies. Insects have many enemies, ranging from bacteria, and fungi to fish, birds and mammals. Against the former they have no defences. Against vertebrates their principal defences are evasion and deception, though more active methods are not rare. There are bugs which emit pungent odours and liquids, ants that liberate formic acid, caterpillars covered with stinging hairs, even butterflies that are seldom eaten because of their unpleasant flavour, and the Bombardier Beetle fires at its enemies a charge of formic acid attended by an audible explosion. Sudden changes of attitude and appearance (terrifying behaviour) are adopted by many insects; for example, the caterpillar of the Elephant Hawk Moth becomes a snake's head, that of the Lobster Moth changes into a giant spider, the Devil's Coach Horse Beetle transforms into a scorpion, the Eyed Hawk Moth opens its wings to display two large staring eyes, and so on. Amongst the more familiar ways of evading enemies is what has come to be known as camouflage: the insect escapes detection by resembling some part of its environment to the point of deception. Good examples are the 'stick' caterpillars of Geometrid moths and the mottled patterns of most moths that rest by day on tree trunks. Light and shade are also made use of but, naturally, all protection is lost the moment the insect moves. Protective resemblance to inanimate objects is indeed widespread in the insect world. Less common is its further development, known as mimicry, in which an unprotected insect comes to resemble some other insect which, perhaps because of its taste or its sting, enjoys a measure of immunity from attack by insectivorous animals. Examples of both these methods of self-defence, evolved by the operation of natural selection, are illustrated on Pages 65 and 66. Of all methods used by insects to avoid their enemies the commonest, naturally, is simply by hiding in nooks and crannies, inside the stems of plants, under rocks and stones, in deliberately constructed cases of silk, leaves and debris, or simply in the soil.

Fascinating as are the protective devices of insects against the attacks of the larger animals—and instances could be multiplied indefinitely—they are of no avail against other insects, which are their most dreaded enemies. In the British Isles some 20,000 different kinds of insects are to be found. Of these about 4,000 are exclusively parasites in or on other insects and at least as many others prey upon other insects in one stage or another. One might with some justification say that one half of the species live at the expense of the other half. Cruel as this is, it is probably true that, were it not so, man's chances of survival in these islands would be decidedly precarious; so high is the rate of reproduction of insects that, were their increase uncontrolled, in two or three years they would destroy all other life.

When considering the animals of the countryside in the British Isles one is prone to think only in terms of birds, a few mammals, frogs, toads, snakes, fishes, snails and so on. It is seldom realised that the insects far outnumber all these both in numbers of species and, even more, in individuals. Taking the world as a whole insects outnumber all other species of animals by at least four to one. They hold in fact a dominant position in the animal kingdom. It is interesting to speculate why this should be so, what special attributes they possess that may have conferred special advantages on them in the struggle for existence. Briefly, these would seem to be their small size, their adaptibility and their power of flight. Their smallness enables large numbers of them to flourish on extremely small quantities of food: a number of minute parastic wasps may successfully reach maturity in a single Hawk Moth egg, a weevil will undergo its whole development in a single grain of wheat; many larvae find all their needs between the two surfaces of a small leaf. Correlated with this is the shortness of their life cycle, which seldom lasts more than one or two years and may be as short as a matter of days. Their adaptibility to their environment is remarkable. No other class of animals has so successfully colonised the land, from pole to pole, as the insects, living in both arctic cold and desert heat, thriving at altitudes that man can barely tolerate, inhabitating petroleum pools and living and breeding in such substances as opium, pepper and strychnine. The tough impermeable cuticle which forms the insect's exoskeleton not only prevents the drying up of the soft internal parts but also provides a covering, both flexible and yet of relatively immense strength; and the wings provide means of dispersal to new areas, of finding food and a mate. It is small wonder that insects have indeed colonised the earth, that every single acre of fertile country holds several millions of them.

THE PRINCIPAL PARTS OF AN INSECT

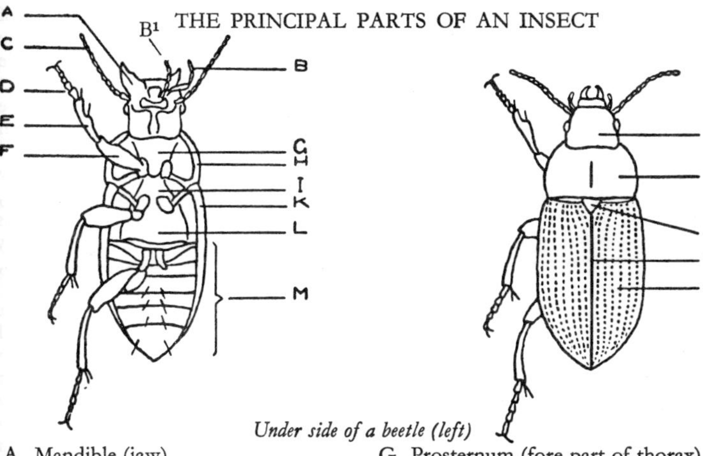

Under side of a beetle (left)

A Mandible (jaw)
B, B[1] Palpi (organs of touch)
C Antenna (feeler)
D Tarsus (foot)
E Tibia (shank)
F Femur (thigh)
G Prosternum (fore part of thorax)
H Edge of thorax
I Mesosternum (middle part of thorax)
K Edge of elytron (wing cover)
L Metasternum (rear part of thorax)
M Abdomen

Upper side of a beetle (right)

1 Head
2 Prothorax (fore part of thorax)
3 Scutellum (only visible part of middle section of thorax)
4 Elytral suture (line of contact of the two elytra)
5 Elytron, showing striae (lines of small pits etc.)

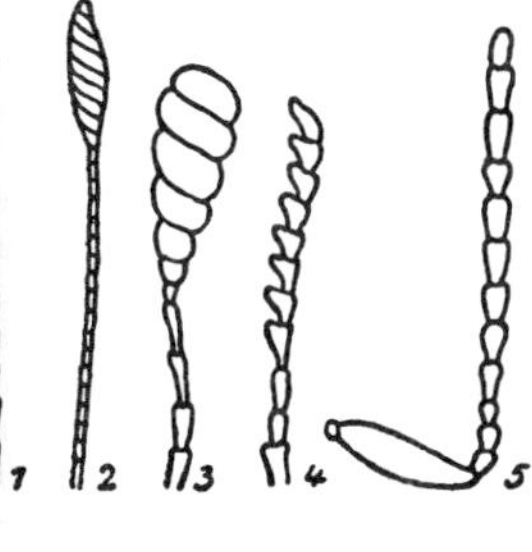

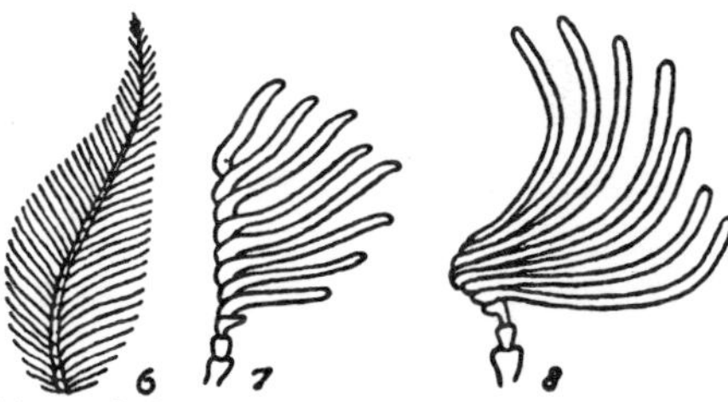

Types of Antennae

1 Filiform (thread like)
2, 3 Capitate (clubbed)
4 Serrate (toothed)
5 Geniculate (elbowed)
6 Bipectinate (feathery)
7 Pectinate (combed)
8 Lamellate (fan-leaved)

CLASSIFICATION

Classification of any group of animals serves two purposes. First, it provides a method of giving convenient and universally acceptable names to all the species, and secondly it shows how these species are related to one another. The animal kingdom is, for these purposes, divided first into a number of *Phyla*, one of which, the Phylum *Arthopoda*, includes all such animals with jointed feet as insects, spiders, mites, crabs, lobsters and shrimps. One of the subdivisions of this Phylum is that containing the true insects. This is the Class *Insecta*. Within this Class there are recognised twenty four major divisions called Orders, the chief characteristics and the names of which follow. The working entomologist, however, is mainly concerned with the next three categories, namely the family, genus and species. The last two provide the scientific name of an insect, such as is used for each insect illustated in this work; and each family is an important group of genera within an Order.

To take an example, the Swallow Tailed Butterfly has the scientific name Papilio machaon. This implies that it belongs to:

the species	Machaon	the order	Lepidoptera
the genus	Papilio	the class	Insecta
the family	Papilionidae	the phylum	Arthropoda

thus, by the combination of two words only, it is given a name different from that of any other animal, and it is placed correctly in the system of classification. The twenty-four Orders of Insects commonly accepted by entomologists are as follows:-

Class Insecta

Sub-class Apterygota (primitive wingless insects)

I Diplura. Bristle-tails
Mostly minute soil-inhabiting insects, eyeless and showing affinities with millipedes. Unlikely to be found except by special collecting methods.
12 British species.

II Thysanura. Bristle-tails
Like Diplura, but with compound eyes. The Silver Fish is a good example of the Order.
11 British species. *Fig.* 1.

III Protura. No popular name
Minute soil insects, about 1 mm long. Eyeless and with piercing mouth parts.
17 British species.

IV Collembola. Spring-tails
Quite small wingless insects sometimes met with in vast swarms. They can always be recognised by having suckers on the underside of the abdomen, just between the legs; and most of them also have under the body a forked springing organ which enables them to jump considerable distances.
261 British species.

Sub-class Pterygota (Winged Insects)

Division 1. Hemimetabola (with incomplete metamorphosis)

V Orthoptera. Crickets, Grasshoppers, Stick Insects etc. These familar insects are well illustrated by *Figs.* 2-7, 262.
38 British species.

VI Isoptera. White Ants. Also called Termites.
No British species.

VII Plecoptera. Sconeflies
32 British species. *Fig.* 11.

VIII Embioptera. Web-spinners
No British species.

IX Dermaptera. Earwigs
All the 9 British species are much like the Common Earwig illustrated in *Fig.* 2.

X Ephemeroptera. Mayflies
The 46 British species are very similar to the Mayfly illustrated (*Fig.* 10) though most of them are much smaller.

XI Odonata. Dragonflies
42 British species. *Figs.* 8-9, 263.

XII Pscoptera. Booklice and allies
These insects are rather suggestive of aphids, and are about the same size. They are most easily distinguished by having biting mouth-parts in place of the sucking beaks of the latter. Many species are winged.
60 British species. See *Fig.* 30.

XIII Anoplura. Lice
286 British species, mostly found on birds. *Fig.* 24.

XIV Thysanoptera. Thrips
Small slender insects with long-fringed wings and stylet-like mouth-parts. Common on flowers.
183 British species.

XV Hemiptera. Bugs
Several samples of these familar insects, characterised by their sucking beaks, are shown in *Figs.* 12-23.
1411 British species.

Division 2. Holometabola (with complete metamorphosis)

XVI Neuroptera. Alderflies, Lacewings etc.
The rather frail translucent net-veined wings are very characteristic of these insects.
54 British species. *Figs.* 25-26.

XVII Mecoptera. Scorpion Flies
Only 4 British species. *Figs.* 28-29.

XVIII Trichoptera. Caddis Flies
Caddis Flies are often mistaken for moths, but can at once be distinguished because they have no scales on the wings and no proboscis.
188 British species. See *Fig.* 27.

XIX Lepidoptera. Butterflies and Moths
2187 British species. *Figs.* 110-193, 265-6.

XX Coleoptera. Beetles
3690 British species. *Figs.* 31-109, 267.

XXI Strepsiptera. Stylops
Very rare minute parasites on bees and some bugs. Males with only the hind wings present; females sack-like internal parasites.
17 British species.

XXII Hymenoptera. Sawflies, Ants, Bees and Wasps, Ichneumon Flies.
6190 British species. *Figs.* 222-252.

XXIII Diptera. Flies (with only two wings)
5200 British species. *Figs.* 194-219.

XXIV Siphonaptera. Fleas
47 British species. *Figs.* 220-221.

DESCRIPTIONS

In the descriptions that follow the first name is the popular (English) name, the second the scientific name. The latter always consists of two words which are either Latin or latinised words. Immediately following them is the name of the author who described the insect and gave it the name quoted. The authors' names are given in full except in the case of Carl von Linne, who originated this binominal system of nomenclature and for whom the simple abbreviation 'L' is always used.

As some of the insects illustrated are much enlarged and others figured are less than natural size, reference should always be made to the text to discover the correct size. In scientific literature measurements are always given in millimetres, and this system is adopted here, except in the case of the butterflies and moths. The sizes for these can easily be converted into millimeters by multiplying by 25. Measurements of length do not include the antennae. The range of each species within the British Isles is only indicated in general terms, as the exact distribution is not yet fully known of many of the species illustrated.

THYSANURA - Bristle-tails

1 Silver Fish, *Lepisma saccharina* L. 10-15 mm.

A rather spindle-shaped insect, tapering towards the tail, which bears three long fine bristles. Pale silver-grey. Very active. Occurs mostly indoors in larders, bathrooms, kitchens and amongst books and papers. Feeds on starchy substances and is sometimes common among stored food products. Widely distributed.

ORTHOPTERA
Earwigs, Cockroaches and Grasshoppers

2 Earwig, *Forficula auricularia* L. 10-14 mm.

This well known and common insect hardly needs description. Though seldom seen flying it does in fact fly well, and when alighting folds its membranous wings fan-wise very quickly and uses the pincers on the hind end of the body to help tuck them into place beneath the short wing-covers. The female 'broods' over her eggs, which is remarkable in so lowly an insect, and cares for the young till they can look after themselves. Usually found in crevices and holes near ground level, and old hollow plant stems. Often damages growing plants but also feeds on small insects. There is no authentic record of the insect entering the ear, as the name suggests. Common throughout the British Isles.

3 Lapp Cockroach, *Ectobius lapponicus* L. 7-11 mm.

Brown, paler along the sides and with a darker mark in the middle of the thorax. The elytra completely covering the wings and body in the male; elytra much shorter and wings almost absent in female. Southern England.

4 German Cockroach, *Blattella germanica* L. 11-13 mm.

Yellowish brown with two dark stripes

on the thorax. Flattened, with long thread-like antenae. Very active. Elytra completely cover the body in both sexes. An introduced species now common in bake-houses, hotel kitchens and centrally heated premises. Sometimes in rubbish tips where conditions are warm enough.
The Common Cockroach *Blatta orientalis* L. or 'Black Beetle' is larger than either of the two species illustrated, broader and much darker in colour with short or very short elytra. Both species are nocturnal.

5 Common Grasshopper, *Chorthippus bicolor* L. 17-20 mm.

General coloration brownish, mottled with lighter and darker shades but very variable. Tip of body generally reddish. Elytra and wings covering the whole body. Very active, flying and leaping. Common throughout British Isles.

6 Wart-biter, *Decticus verrucivorus* L. 35-45 mm from head to wing tip.

Green with brownish spots on elytra and hind legs. Hind legs almost twice as long as body. Rare in Britain, and only in southern counties.

7 Great Green Grasshopper, *Tettigonia viridissima* L. 45-55 mm.

Clear green without markings. Antennae long and slender. Elytra extending beyond body. Sings loudly, often far into the night, by rubbing elytra together. The largest British grasshopper, widespread in southern counties especially in coastal districts.

ODONATA - Dragonflies

8 Brown Aeshna, *Aeshna grandis* L. Length from head to tail 70-75 mm.

Brown with yellow marks on side of thorax and small blue or yellow spots on abdomen. Wings faintly brownish and held flat when at rest. Swift and competent flyer, catching prey on the wing. The larva (or nymph) lives in still water and is exceedingly rapacious, feeding on insects, and even very small fish, which it catches with its extensile pincers or 'mask'. When the nymph is fully fed and the dragonfly due to emerge from it, it climbs out of the water onto the stem of a water plant. After the dragonfly has emerged and flown away, the empty nymph-case is left clinging to the plant. Often found flying far from any water. Common in Southern England, rare elsewhere.

9 The Demoiselle, *Agrion virgo* L. Length from head to tail 45 mm.

Head and body of male metallic blue; of female metallic green. Wings of male iridescent bluish brown; of female pale brown. A slow flyer and rests with wings closed over the back. Always flies over or very near water. Nymph in running water. Common locally in southern England, Wales and Ireland.

EPHEMEROPTERA - Mayflies

10 Mayfly, *Ephemera vulgata* L. Wing spread 32-38 mm.

Dull yellow to pale brown; wings glistening with a dense network of veins. Three long filamentous tails. Mouth parts atrophied so that the mayfly takes no food. It mates in the air during evening swarms and lives only a few hours. Eggs dropped into water. Nymph rather like adult but with gills and no wings. When full fed, after about a year, climbs out of water and changes into a winged pre-adult which moults yet again before being able to fly.

PLECOPTERA - Stoneflies

11 Stonefly, *Perla cephalotes* Curtis
Wing span 50-55 mm.

Brown, wings held flat over back when at rest. Nymph only in running water, under stones or amongst gravel, and having only two filamentous tails. Adult sluggish, always near water.

HEMIPTERA - Bugs

12 Water Scorpion, *Nepa cinerea* L.
Length, including breathing siphon, about 35 mm.

Predacious, the first pair of legs serving as pincers. Under the wing-covers (elytra) are purplish wings, but the insect does not fly as it has no wing muscles. Lives entirely in rather shallow muddy waters. Throughout British Isles, but rare in north.

13 Water Boatman,
Notonecta glauca L. Length 15 mm.

Hindlegs modified to form excellent 'oars' fringed with long hairs, and the back 'keeled' as the insect always swims on its back. Dives and flies readily and is carnivorous. Common throughout the British Isles.

14 Great Pond Skater, *Gerris naias* Degeer. Length 13-17 mm.

The front legs are used for seizing prey, the other two pairs for 'skating' over the surface of the water in a characteristic jerky fashion. Dives very rapidly. Wingless.
Throughout the British Isles, except Scotland.

15 Lesser Water Boatman,
Corixa punctata Illiger
Length 10-15 mm.

Dark greenish or blackish brown, darker than the Water Boatman; not keeled on back and not swimming on its back, and more often below the surface. Flies by night and feeds on diatoms, algae etc. Throughout the British Isles.

16 Bed Bug, *Cimex lectularius* L.
Length 5 mm.

Wingless in both sexes and all stages, and very flat except after feeding. Blood-sucking and almost exclusive to man. Has faint, unpleasant odour. Hides by day in crevices of woodwork, behind wall paper and in similar places. Formerly common throughout the British Isles, but the use of modern insecticides, especially D.D.T., has greatly reduced its numbers.

17 Beadle or Soldier Bug,
Pyrrhocoris apterus L.
Length 8-9 mm.

Occurs in both winged and wingless forms, the former exceedingly rare in England. Very rare in southern England, but sometimes abundant locally. Attached to mallow.

18 Toothed Shield Bug,
Pictomerus bidens L.
Length 10-12 mm.

The name refers to the pointed shoulders. Found on most trees and shrubs in rather damp places, where it feeds voraciously on other insects, even caterpillars larger than itself. Found throughout the British Isles.

19 Blood-red Shield Bug,
Acanthosoma haemorrhoidale L.
Length 12 mm.

Pale shining brown. It takes its name from the bright red bands on its body, which are conspicuous when it takes to flight. Occurs on many trees and shrubs, but especially hawthorn. In Britain it occurs in the southern counties only.

20 Cabbage Shield Bug,
Eurydema oleraceum L.
Length 6-7 mm.

Green with markings which may be either red or whitish. Feeds mostly on low-growing plants of the Cabbage family, but in the British Isles is never common enough to do any damage. In Britain, it is found in the southern counties only.

21 Alder Frog-hopper,
Aphrophora alni L. Length 12 mm.

One of a group of several species of Frog-hoppers all of which, when nymphs, produce the well-known cuckoo-spit found on many plants. This species occurs chiefly on alder and sallow. The frothy cuckoo-spit is derived from surplus fluid sucked out of plants by the nymphs and discharged from the tip of the body. Generally common.

22 Mealy Plum Aphis,
Hyalopterus arundinis Fabricius
Length 2 mm.

A very distinct *Aphis* which is characterised by its mealy covering. Attacks plum trees and causes the leaves to curl. Its alternate host plants are water grasses and rushes to which the winged individuals fly. Has many generations during the year, some of which reproduce parthenogenetically ('virgin birth'), others after pairing. The last autumn generation at least consists of both males and females, winged, which return to the plum trees and there lay over-wintering eggs. A five per cent tar-distillate winter wash applied in December or January is the best antidote.

23 Scurvy Scale,
Aulacaspis rosae Bouché
Length 15-25 mm.

The stems of roses are sometimes so infested with this scale insect as to appear whitish. The rounder scales are those of the females, the longer narrower ones those of the male. Males with wings appear in early summer. After mating, both sexes settle on the rose stems from which they suck the juice, and quickly secret the material which hardens to form their protective coverings.

ANOPLURA - Lice

24 Head Louse, *Pediculus humanus* L.
Length 1-2 mm.

A dirty grey-coloured insect that can only live on man. Has two distinct strains one of which lives among the hair of the head, the other on the body. Its eggs are known as nits and are attached to hair, to which the insect's claws are suitably adapted for clinging. To feed, it punctures the skin of the host and sucks blood. Is a great danger to health as it is capable of transmitting such serious diseases as typhus. Easily controlled by the use of D.D.T. powders.

NEUROPTERA - Lacewings

25 Green Lacewing, *Chrysopa carnea* Stephens. Wingspread 25-30 mm.

The translucent green wings and golden eyes are very characteristic. Both larvae and adults feed chiefly on Aphids. The eggs are attached to the tops of long thin stalks, to escape being eaten by other insects. The operations of making the stalk and laying the egg on the top of it are performed all in one movement. The fly is common everywhere.

26 Giant Lacewing,
Osmylus fulvicephalus Scopoli
Wingspread 44-48 mm.

The head is orange the eyes greenish-black. There are a few pale markings on the black thorax and abdomen. Wings hyaline with brown spots. The larvae live in wet moss, feeding on the grubs of small flies in the mud. The adults occur amongst dense vegetation along woodland streams in England and Wales, Ireland and Scotland but are local and rare.

TRICHOPTERA - Caddis Flies

27 Great Red Sedge,
Phryganea grandis L.
Wing spread 35-60 mm.

This is the largest British Caddis Fly, and its English name is that used by fishermen. It flies at dusk mainly and is very much like a moth, but has no scales on its wings. The forewings are narrower but tougher than the hind wings. Its grub lives in water in a case made of fragments of sticks and stems of plants, feeding on other small insects, worms etc. It is a common insect, found near water throughout the British Isles.

MECOPTERA - Scorpion Flies

28 Snake Fly,
Raphidia notata Fabricius
Wingspread 25 mm.

This curious black-bodied insect derives its name from the way it moves when in search of prey. The long neck is really the first segment of the thorax, not a true neck, and is very flexible. Its grub lives under the loose bark of trees and in rotting wood, feeding on soft-bodied creatures that inhabit such places. It is not common, and in England occurs only in the southern counties.

29 Scorpion Fly,
Panorpa communis L.
Wingspread 25-30 mm.

When disturbed, the male curls the tip of the abdomen up like a scorpion; but it is quite harmless. It feeds on small dead insects, its pointed head, with the mouth at the tip, being well suited for the purpose. Its grub lives in the soil, and it is a common insect throughout the British Isles.

PSOCOPTERA - Booklice, Psocids

30 Book Louse,
Liposcelis divinatorius Mueller
Length 1.5 mm.

Its colour is light brown, and it is entirely wingless. Though often found running about among old papers, amongst insect collections and in similar places, it is not really a louse. It feeds on minute moulds and mildews, and its presence indicates dampness and lack of ventilation. An allied species can make a faint noise by tapping with its body and hence is sometimes called the 'Death Watch'.

COLEOPTERA - Beetles

31 Tiger Beetle,
Cicindela campestris L. 12-16 mm.

An extremely active and very beautiful beetle. As rapid in flight and in taking to flight, as a bluebottle, for which it is easily mistaken on the wing. Its colours are best seen when slightly magnified. It has a faint, rather agreeable odour, and is highly predacious in all stages. Found in warm sandy banks and such places through the British Isles where its grub inhabits burrows about a foot deep.

32 Woodland Ground Beetle,
Carabus nemoralis Mueller
22-26 mm.

One of the larger common ground beetles to be found in the country and in gardens and parks in towns. The thorax is purplish, the elytra more brassy or coppery. Nocturnal, hiding by day under stones and debris of all kinds. Predacious in all stages. Throughout England, Ireland, Wales and and Scottish lowlands.

33 Burrowing Ground Beetle, *Clivina fossor* L. 55-65 mm.

A rather small ground beetle, the thorax markedly separated from the elytra. Forelegs shaped for digging. Frequents damp situations where it burrows in the mud; rarely found in the open, but generally common from Scottish lowlands southward.

34 Shiny Ground Beetle, *Harpalus aeneus* F. 8.5-10.5 mm.

The elytra vary from black to brassy, and may be purplish, green or bluish; the legs and antennae are red. Abundant in fields, preferring rather dry places, under stones, etc. throughout Britain.

35 Giant Water Beetle, *Dytiscus marginalis* L. 30-35 mm.

Beautifully shaped for swimming rapidly in water, the hind legs in particular being shaped rather like oars, fringed with long hairs. The wing covers (elytra) of the male are smooth, those of the female deeply furrowed. Part of the foreleg of the male is shaped to form a 'sucker' with which the female is grasped during mating. Both larva and adult are highly predacious, and the latter is a strong flyer when out of the water. Throughout British Isles.

36 Whirligig, *Gyrinus substriatus* Stephens 5-7 mm.

These extremely wary and active beetles are very aptly named. They swim endlessly round and round on the surface of streams and pools usually in groups looking like gyrating drops of molten metal. The middle and hind legs are especially adapted for swimming, the front pair being long and slender. The eyes are divided, the upper half to see above the water surface, the other below. Very common in southern England especially.

37 Black Rove Beetle, *Stenus biguttatus* L. 4.5-5 mm.

This species and the next are representives of the great family of Rove Beetles of which nearly four hundred different kinds are found in the British Isles alone. They are very easily recognised by their narrow shape and very short elytra. A local species found chiefly in sandy places by streams and shores from southern Scotland southwards. The best known species in the family is the big jet-black 'Devil's Coach Horse'.

38 Red-backed Rove Beetle, *Staphylinus caesareus* Cederhjelm 14-20 mm.

This species may sometimes be seen settling on hot pathways in the spring; also to be found under stones, clods of earth etc. Its red elytra are very conspicuous. Scottish Lowlands southwards.

39 Ant-nest Beetle, *Claviger testaceus* Preyssler 2-2.5 mm.

This oddly shaped beetle is very small and uniformly yellow. It lives as a guest in the nest of the small yellow ants *(Lasius)* which are found under stones and in similar places, and feeds largely on their grubs. Mostly in chalky districts in Southern England.

40 Glow-worm, *Lampyris noctiluca* L. 11-12 mm; female 12-18 mm.

The male looks rather like a small 'Sailor Beetle', is fully winged and has full sized elytra. The much bigger female looks more like a beetle grub than a fully grown beetle. All stages feed upon snails and all are more or less luminous: but it is especially the full-grown female that is the well known glow-worm to be seen often in damp ditches and hedge sides, the light being produced mainly from underneath the tip of the abdomen. England to the Scottish lowlands, local.

41 Sailor Beetle, *Cantharis fusca* L.
11-15 mm.

This common and active beetle flies freely in the sun. It has close relatives, in which the wing covers are brown, known as Soldier Beetles. All are predacious in all stages, and will even attack each other, although their wing cases and horny exterior are rather soft in comparison with most beetles. Southern England.

42 Ant Beetle,
Thanasimus formicarius L.
7-10 mm.

So-called because of its ant like appearance and behaviour. Not actually associated with ants. Both as grub and beetle it attacks bark-boring beetles, and is usually found running about on fallen timber and logs. England to southern Scotland, but local.

43 Brassy Pollen Beetle,
Meligethes aeneus F. 1.5-2.7 mm.

There are many species, even in the British Isles, of these small Pollen Beetles, and because in all stages they live on flowers, some of them are troublesome. They are very difficult to distinguish, but in all of them the elytra leave the extreme tip of the abdomen exposed. Common throughout Britain.

44 Seven-spot Ladybird,
Coccinella septempunctata L.
5.5-7.5 mm.

This is the largest British ladybird beetle, widely distributed in all kinds of country. In all stages it feeds on aphids and is a most useful insect.

45 False Ladybird,
Chilocorus renipustulatus Rossi
4-5 mm.

Might be mistaken for a variety, of which there are many, of the very common Two-spot Ladybird, with the colours reversed; but it is rounder. Like the true ladybirds, it feeds on aphids, and is a beneficial insect. Southern England.

46 Bacon Beetle,
Dermestes lardarius L. 7-9 mm.

Modern hygiene and storage methods have rendered this insect much less common than it used to be. It is by no means confined to bacon as a food and will eat almost any dead dry animal matter, especially hides. When disturbed it folds in its legs and antennae and presents a very perfect oval outline.

47 Fur Beetle, *Attagenus pellio* L.
4-5.5 mm.

A close relative of the last species but, as the name implies, has a preference for furs. It will also attack woollen goods and may do serious damage in the home, and especially in warehouses where the goods are not frequently disturbed. Throughout Britain.

48 Common Burying Beetle,
Necrophorus vespillo L. 12-22 mm.

Both this species and the equally common *Necrophorus investigator* have two red bands across the elytra, but whereas the front of the thorax in *N. vespillo* has patches of downy hair, in *N. investigator* it is bare. England and Ireland, rarer in Scotland.

49 Black Burying Beetle,
Necrophorus humator Goeze
18-25 mm.

This Beetle is easily distinguished from all the other Burying Beetles by being entirely black except for the tips of its antennae, which are red. It is widely distributed throughout Britain and often common.

50 Red-necked Sexton,
Oeceoptoma thoracicum L.
12-16 mm.

It is not really the neck but the thorax that is red in this beetle, which is a rather flattened insect with distinct raised ridges running the length of its elytra. In carcases, fungi etc. throughout Britain.

51 Nature's Scavengers

Any rotting carcase will attract burying beetles during the spring and summer when they are on the wing. Having found a suitable dead mouse or bird, for example, the beetles dig away the earth from underneath it so that the carcase gradually sinks deeper and deeper into the ground, the excavated earth being used to cover it on top. When it is buried, the beetles, which themselves feed on it, lay their eggs on it in the certainty that it will supply ample foodstuffs for their larvae when they hatch.

52 Lined Click-beetle,
Agriotes lineatus L. 7.5-10 mm.

This narrow, dark brown beetle is less well known than its thin light brown wiry grub, which is the common wireworm that does immense damage to the roots of grass, corn and many other crops. There are several other closely related species. When occurring together in the soil they may build up a population totalling many millions per acre. Throughout Britain.

53 Scarlet Click-beetle,
Elater sanguineus L. 12-17,5 mm.

One of the handsomest of the British Click-beetles, and especially associated with oak woods in the south of England. Click-beetles, or Skipjacks as they are sometimes called, are so-called from their habit of springing into the air with a clicking sound when turned on to their backs on the ground. Southern England, but very rare.

54 Comb-horned Upland Click-beetle, *Corymbites pectinicornis* L.
12-13 mm.

This very beautiful Skipjack is well described by its popular name, for its antenna when opened out are very comblike. It prefers hilly uplands, where its 'wire-worm' larva feeds on grass roots in pastureland. It is a rather rare and local species, found in Britain in the Midlands and northward to Scotland.

55 Spider Beetle, *Ptinus fur* L.
2-4 mm.

This beetle is characteristic of a whole group of species, some of them much more spider-like in shape than the one illustrated. They curl up into a ball when they sham death. All are very liable to be troublesome in warehouses and wherever foodstuffs are stored for they and their larvae feed particularly on dead and dry animal and vegetable matter. Generally distributed.

56 Furniture Beetle,
Anobium punctatum De Geer
3-4 mm.

Known in the furniture trade as 'worm', this beetle makes the characteristic holes in old furniture, and often in new wood. The grub tunnels inside the wood, and produces quantities of fine powder which sometimes falls to the ground in severe attacks. The holes are made by the fully grown beetle when it emerges from the wood (when the damage is done!) to find a mate and so to repeat the process. The eggs are laid in tiny cracks on the surface of the wood. A common insect in woodlands, as well as in furniture.

57 Oil Beetle, *Meloe proscarabaeus* L. 13-32 mm.

This wingless and rather ungainly beetle is chiefly interesting on account of its curious life history. When handled it exudes a clear yellow oil from its joints. The females lay enormous batches of minute eggs in small holes in the ground. The tiny grubs when they hatch are exceedingly active and if possible find their way into flowerheads that are visited by bees, for their sole chance of survival rests upon their attaching themselves to a bee and so being carried to its nest. Once in the nest the grub changes to a sluggish creature which proceeds to devour the food stored up by the bee for its own young. Widely distributed in southern England, rare northwards.

58 Cellar Beetle, *Blaps mortisaga* L. 20-30 mm.

This black, slow-moving beetle is aptly named. It is assocated with dark damp places such as cellars, old-fashioned sculleries, stables and outbuildings. The one illustrated is a rare species only found occasionally in the north of England. Like the similar and much commoner southern species, *Blaps mucronata,* it has a distinctly unpleasant smell.

59 Mealworm Beetle, *Tenebrio molitor* L. 14-16 mm.

This insect is much less familiar than its grub, which is the well known mealworm commonly bred in quantities as food for small birds and other insectivorous animals. It can be very troublesome when infesting warehouse stores of grain or flour.

60 The Tanner, *Prionus coriarius* L. 24-40 mm.

The thorax of this handsome 'Long-horn' has three short spines on each side. The adult flies chiefly at night. The larva bores in the wood, especially in the roots, of oak, beech, birch and pine growing in rather damp situations, doing considerable harm to the trees. Southern England and Wales to Midlands.

61 Red-brown Long-horn Beetle, *Leptura rubra* L. 12-18 mm.

The sexes of this brightly coloured longicorn, as shewn in the illustrations, are different in coloration. Its larvae are wood-borers, like those of most long-horn beetles. It attacks exclusively coniferous trees. Occurs sporadically, possibly as an 'escape' from imported timber.

62 Four-banded Long-horn Beetle, *Strangalia quadrifasciata* L. 13-18 mm.

This handsome longicorn occurs rarely in southern counties at elder blossom and some other flowers in summer. Its larva feeds in the dead and decaying wood of oak, alder, poplar and birch, especially old stumps in moist situations. Rarer northwards.

63 Short-winged Long-horn Beetle, *Molorchus minor* L. 8-13 mm.

Though small, this beetle is conspicuous for its very long antennae and much shortened elytra. Its larva burrows principally in broken branches of pine, larch and spruce and in the stacked wood of these trees. Southern England; local.

64 Violet Long-horn Beetle,
Callidium violaceum L. 10-15 mm.

This beetle has a blue and violet sheen. Formerly it was a great rarity in Britain but is now much commoner. Its larva burrows between the bark and the sap-wood of felled coniferous trees such as larch, spruce and pine, often doing much damage to palings, posts and rustic work that have not been stripped of their bark. Southern counties to Midlands, rare, possibly not indigenous.

65 Firewood Long-horn Beetle,
Spondylis buprestoides L.
15-24 mm.

Has been found only very rarely in Britain in timber imported from the Continent. Its larva burrows in various coniferous trees.

66 Dusky Long-horn Beetle,
Criocephalus rusticus L. 12-16 mm.

Not so common in Britain as the very similar *Criocephalus ferus*. Both have the similar habit of attacking the stumps, boles and roots of pine and larches recently killed or previously attacked by other wood-boring beetles.

67 Chestnut Long-horn Beetle,
Tetropium castaneum L. 10-18 mm.

The beetle illustrated is found occasionally in imported timber. The very similar *T. gabrieli* Weise occurs in many pine, larch and spruce plantations, where its larva lives just under the bark and in the outer sapwood of sickly trees, seldom attacking healthy ones.

68 Musk Beetle, *Aromia moschata* L.
18-32 mm.

The scent of this Long-horn is so strong that it can often be noticed at some distance from the trees which it affects, principally old willows and sallows in which the larvae bore. The beetles are usually green, but may be bluish, and fly freely in the sun in June and July and in some places in southern England are quite common; extends to Ireland and Scottish Lowlands.

69 House Long-horn Beetle,
Hylotrupes bajulus L. 10-20 mm.

Attacks dry seasoned wood of coniferous trees such as the roof timbers of houses, posts and dead trees. Formerly rather a rarity, it has become a serious pest in some outlying parts of Greater London. Its larva may bore in timbers for several years without disclosing its presence, completely hollowing them. Control is very difficult, and prevention of attack by treatment of timbers, before or during erection is thus recommended. The blackish beetle flies freely in hot, sunny weather.

70 Large Poplar Long-horn,
Saperda carcharias L. 18-28 mm.

This is another Long-horn that attacks poplar trees. Its grubs bore into the heart of the tree and thus may render the timber commercially worthless. The adult female lays her eggs usually near the base of young trees; if this is protected attack may be prevented. Southern England to the Midlands; scarce.

71 Timberman,
Acanthocinus aedilis L. 12-20 mm.

The astonishingly long antennae of this beetle, especially in the male, are its most remarkable feature. Apart from casual emergences from imported timbers, such as pit props, it is found in old long-established pine woods, such as the Black Forest at Rannoch, where it may be seen flying freely and settling on old pine logs and tree stumps. The males are decidedly pugnacious. The larvae excavate wide galleries just under the bark, seldom damaging the wood of the tree except when making their pupal chambers.

72 Pine-bark Long-horn,
Rhagium inquisitor L. 12-18 mm.

This Long-horn is rather similar to the Timberman but smaller, with much shorter antennae. It is commoner, is quite active and has a very characteristic 'watchful' resting attitude. The larvae live under the bark of conifers, excavating galleries on the surface of the wood; they prefer dead or dying timber. Found in Scotland and northern England.

73 Water Jewels, (*Donacia* F.)

73a *Donacia aquatica* L. 6-10 mm.

73b *Donacia clavipes* F. 7-12 mm.

73c *Donacia cinerea* Herbst
7-11 mm.

73d *Donacia crassipes* F. 9-13 mm.

The illustrations give but a poor idea of the beauty of these beetles which, when seen alive on a water lily leaf among rippling water, are truly living gems. They are all attached to water plants, such as reeds and sedges of various kinds, in the stems in which the larvae live. All are rather local, though sometimes numerous where they occur. Most frequent in southern counties of England.

74 Scarlet Lily Beetle,
Lilioceris lilii Scopoli 6-8 mm.

The vivid scarlet of this beetle soon fades after death. It sometimes damages ornamental lilies, on the leaves of which it lays its eggs, glued together. The larva has the curious habit of covering its back with its own dried faeces as a protection, possibly, against its enemies and the sun. Rare, in southern England.

75 Four-spot Willow Beetle,
Phytodecta viminalis L. 5.5-7 mm.

The black spots of this brick red beetle vary considerably in size, shape and number. It is commonly to be found on willow and sallow, on the leaves of which it feeds. Southern counties of England to Durham, local.

76 Alder Leaf Beetle,
Agelastica alni L. 5-7 mm.

This rather metallic coloured beetle is sometimes found on alders the leaves of which it riddles so as to cause considerable damage. Rare in England.

77 Sallow Leaf Beetle,
Lochmaea capreae L. 4-6 mm.

A small brownish beetle which devours the leaves of sallow, particularly of young trees. Southern England to the Midlands. The Scottish Heather Beetle, *Lochmaea suturalis* Thomson, is very similar.

78 White Deadnettle Beetle,
Chrysomela fastuosa Scopli
5-7 mm.

Golden green, the thorax and a stripe down the back and on each elytron blue or violet. Feeds on various nettles. Local but widely distributed.

79 Red Poplar Leaf Beetle,
Melasoma populi L. 9-11 mm.

The elytra red, otherwise blue-black or greenish, metallic. On young poplars, aspens and dwarf sallows. Locally common in England; rare in Scotland.

80 Tansy Leaf Beetle, *Galeruca tanaceti* L. 6-10 mm.

Entirely black, dull or shining. Principally on tansy, but also on other plants. Throughout the British Isles, but not common.

81 Turnip Flea Beetle, *Phyllotreta nemorum* L. 2.5-3.5 mm.

Its jumping habit and fondness for turnips give this tiny beetle its popular name. It sometimes does enormous damage by stopping the growth of the young plants. Occurs throughout the British Isles, in several broods each year. Modern insecticides keep it in check.

82 Cabbage Leaf Beetle, *Haltica oleracea* L. 3-4 mm.

Generally metallic, shining green, but may be bluish. Found principally on *Brassica* (Cabbage) but also on some other plants, mostly in southern England.

83 Tortoise Beetle, *Cassida nebulosa* L. 5-7 mm.

May be rust brown or greenish in colour, and is found chiefly on clover, beetroot and other low plants. Southern counties; rare.

84 Vine Weevil, *Otiorrhynchus sulcatus* F. 9-13 mm.

General colour rather shining black, with patches of yellowish hairs. Lives on roots of plants etc. and may be injurious to vines (in countries where these are grown), strawberries etc. Midlands and southern England; rarely to southern Scotland.

85 Silver-green Leaf Weevil, *Phyllobius argentatus* L. 5-6 mm.

The green colour is due to a covering of scales, mixed with hairs. Common throughout the British Isles, on young birch, oak and other trees.

86 Pea Weevil, *Sitona lineatus* L. 3.5-5 mm.

Ground colour black, but covered with alternating lines of light and darker scales. Abundant throughout the British Isles, and from early spring to late autumn, on clover, peas, vetches and allied plants.

87 Pine Weevil, *Hylobius abietis* L. 9-14 mm.

Pitch-black with patches of yellow scales. Common everywhere on pines and firs during summer months.

88 Banded Pine Weevil, *Pissodes pini* L. 6-9 mm.

The band of yellow scales is characteristic. Found only in the northern counties and Scotland, rather rarely, on pines and fir trees.

89 Grain Weevil, *Calandra granariae* L. 2.0-3.5 mm.

Only in granaries, bakeries and such places, where it is sometimes exceedingly destructive, feeding on stored grain. It is in fact *the* Weevil. The whole development, from egg to beetle, takes place within a single grain.

90 Nut Weevil, *Balaninus nucum* L. 5-8 mm.

On hazel, chiefly in southern England and the Midlands. Eggs are laid on the unripe nuts and the larvae feed in the kernels.

91 Dark Blue Pine Weevil, *Magdalis violacea* L. 4-6 mm.

Larva on small branches of pine trees, of which the beetle damages the new shoots. Only in Scotland, and very rare.

92 Apple Blossom Weevil, *Anthonomus pomorum* L. 3.4-4.5 mm.

Attacks the buds and flowers of apple and pear trees and often does great damage. Throughout the British Isles as far as southern Scotland.

93 Clover Weevil, *Apion apricans* Herbst 2.5-3 mm.

Feeds on clover and is liable to occur wherever this grows wild or as a crop. Sometimes does considerable damage, skeletonising the leaves.

94 Birch Leaf Roller, *Bycticus betulae* L. 6-8 mm.

The female rolls the leaves of birch into characteristic tubes about two inches long and lays its eggs inside. The leaf rolls are to be seen also on hazel and poplar in southern England in spring and early summer; much rarer further north and in Scotland.

95 Bark Beetle, *Crypturgus pusillus* Gyllenhall 1-1,5 mm.

Lives in the galleries formed by other and larger Bark Beetles in pine, larch and other coniferous trees and does not appear to be of any economic importance itself. Southern England, local.

96 Pine Shoot Beetle, *Myelophilus piniperda* L. 4-4.5 mm.

Beetle bores into tender pine shoots and eats its way along the centre, causing the tip eventually to break off; larvae in burrows under the bark of live or decaying trees. Widespread, especially in the north, and causing considerable damage.

97 The Pattern Maker, *Ips typographus* L. 4.5-5.5 mm.

This bark beetle attacks many kinds of coniferous trees, but fortunately is not very common in Britain. The main tunnel is made by the parent female, the branch tunnels by the larvae. In a hot season there may be two or even three broods in a year.

98 Stag Beetle, *Lucanus cervus* L. 30-80 mm.

The male Stag Beetle is a not uncommon sight flying slowly high in the air over gardens in the London area in the evenings about June. The female is less often seen. The grub lives mainly in the stumps of old trees, especially oaks, and takes several years to reach full growth. Common in some southern counties; more rarely as far north as the Midlands.

99 Lesser Stag Beetle, *Sinodendron cylindricus* L. 12-16 mm.

This small Stag Beetle has only a single horn, which curves upwards and backwards and bears long yellow hair to the rear. The female has only a trace of this horn. Flies at dusk and is distributed throughout the British Isles, though commonest in the southern counties. The grub lives in the rotting stumps of ash, beech, willow and other trees.

100 Dung Beetle, *Aphodius fimetarius* L. 5.5-7 mm.

One of a number of very similar and closely related species all of which live and breed in dung. Common throughout Britain.

101 Dor Beetle, *Geotrupes stercorarius* L. 16-24 mm.

Flies with a slow humming flight on warm evenings in summer and autumn. Larvae feed on dung in chambers dug by females to a length of a foot or more below the dung and provisioned with dung carried down from above. Occurs throughout the British Isles.

102 Summer Chafer, *Amphimallus solstitialis* L. 14-18 mm.

Flies about hedges and trees at dusk, sometimes abundant locally, in the south and west of Britain. Larva feeds on roots of small trees and bushes, and at times is destructive to the roots of corn and grasses.

103 Cockchafer, *Melolontha melolontha* L. 20-30 mm.

Known also as the 'June Bug' this beetle is a very common sight in southern counties. Its 'bumbling' flight often brings it into houses. Its grub lives in the soil, feeding for three years on all kinds of roots, and is a pest to the gardener and agriculturist. Rather rare in northern England and Scotland.

104 Garden Chafer, *Phyllopertha horticola* L. 8-12 mm.

Like the Cockchafer this smaller chafer is an important pest, the grubs often doing extensive damage to turf by eating and destroying the roots. The beetle itself also damages the foliage of trees. Common throughout Britain.

105 One-horned Dung Beetle, *Copris lunaris* L. 25 mm.

The closest British relation of the Scarab Beetle of Egypt, and like it, feeding on and breeding in dung. Sandy places in southern England, rather local.

106 Three-horned Dor Beetle, *Typhaeus typhaeus* L. 12 mm.

The males vary a good deal in the size of the horns. The female lacks them. Found chiefly under rabbit droppings; rather local and not found north of the Midlands.

107 Bronze Dung Beetle, *Onthophagus vacca* L. 1.5 mm.

To be found locally only from Kent to Somerset, in dung. The female buries pellets of dung rather deeply in the soil, for the larvae to feed on. Like all dung-beetles, a very useful scavenger.

108 Rose Chafer, *Cetonia aurata* L. 14-22 mm.

This very lovely beetle has its elytra so shaped that they can remain closed whilst the wings are in use. Its popular name is due to the fact that it feeds freely on the foliage of roses, as also indeed on that of many other shrubs and trees. The grubs live for two or three years in the soil, sometimes in ant's nests. Local in southern England becoming rarer northwards, and very rare in Scotland.

109 Bee Chafer, *Trichius fasciatus* L. 9-13 mm.

Due to its covering of hair, this beetle has a rather downy appearance. A northern insect, more often met with in Scotland than in England, but not common anywhere. The grub lives in rotting stumps of birch, alder and other trees.

LEPIDOPTERA

Butterflies and Moths

110 Swallowtail, *Papilio machaon* L. Expanse 3-3½ inches.

Confined in Great Britain to the fens and Norfolk Broads, where the caterpillar feeds on fennel. Just behind its head the caterpillar (shown in the figure) has a glandular defensive organ called the osmeterium which it thrusts out when disturbed. Flies in June; sometimes again in August.

111 Marbled White,
Melanargia galathea L.
Expanse 2 inches.

Not a true 'White' but related to the Meadow Brown. Flies in grassy places, especially on chalk downs in southern England, in late July or early August. Caterpillar feeds on Cock's foot and Timothy Grass.

112 Large White, *Pieris brassicae* L.
Expanse 2½ inches.

So common some years as to do great damage to cabbage and related crops, hence often called the Cabbage White. A migrant, which only survives the winter in small numbers in favourable situations. In a fine season there may be two or three broods. Kept under control naturally to a large extent by a tiny parasitic wasp, *Apanteles glomeratus*.

113 Green-veined White,
Pieris napi L. Expanse
1½-2 inches.

Common throughout the British Isles except the extreme north. The spring brood, which flies in May is smaller, and duskier that the summer brood of August, and the males of both have only one black spot on the forewing. Caterpillar feeds on many of the cresses, rape, horse-radish etc.

114 Brimstone, *Gonepteryx rhamni* L.
Expanse 2¼ inches.

Hatches out of the chrysalis in late summer, hibernates throughout the winter and takes to wing again in early spring, flying till June. Caterpillar feeds on buckthorn. Widespread in England and Wales; local and rather rare in Ireland.

115 Black-veined White,
Aporia crataegi L.
Expanse 2½ inches.

Now extinct in the British Isles. Formerly it was common in most southern counties, and sometimes troublesome in orchards. The young caterpillars are gregarious and live in a shelter of silken webs on the leaves of sloe, hawthorn, plum etc.

116 Orange Tip,
Anthocharis cardamines L.
Expanse 1½ inches.

The female lacks the orange patch which adorns the forewing of the male. Common in England and Wales, but rare in Scotland. One of the first butterflies to emerge in spring, in May. Caterpillar feeds on many Cruciferae such as Lady's Smock, hedge mustard and charlock.

117 Clouded Yellow,
Colias crocea Geoffroy
Expanse 2 inches.

Only the male is illustrated. The female is less brilliant and has yellow spots in the black borders and has two colour forms; one is like the male, the other is very pale yellow and is the *helice* form. Very seldom survives the English winter; those seen in early summer are migrants which, in a favourable season give rise to a second brood in August and September.

118 White Admiral,
Limenitis camilla L.
Expanse 2¼ inches.

Flies in July in most large woodland areas in southern England, with a gliding flight. Caterpillar feeds on honeysuckle, preferring shade, and hibernates when very small. It was originally called the White Admirable.

119 Red Admiral, *Vanessa atalanta* L. Expanse 2½ inches.

A common butterfly in most seasons, but it does not regularly survive the winter. Reinforced annually by migrants from the continent, the first of which arrive in June. Most numerous in the autumn, when it is attracted by over-ripe fruit. Caterpillars feed on Stinging Nettles, each one constructing a kind of tent of leaves forming a shelter.

120 Peacock, *Nymphalis io* L. Expanse 2½ inches.

The first Peacocks usually appear on the wing in August, and are especially attracted to the flowers of Buddleia. Later, they visit over-ripe fruit and Michaelmas Daisies freely. They then go into hibernation till the following spring, and may often be seen till early May. The caterpillars feed gregariously on Stinging Nettles.

121 Small Tortoiseshell, *Aglais urticae* L. Expanse 2 inches

One of the commonest British butterflies. Those that appear in the autumn hibernate in barns and outhouses, reappear in the spring, and give rise to a summer brood about July. The caterpillars feed gregariously on Stinging Nettles. Occurs throughout the British Isles.

122 Camberwell Beauty, *Nymphalis antiopa* L. Expanse 2½-3 inches.

Occurs in the British Isles only as a very rare migrant believed to come from Scandinavia, possibly transported accidentally by ships importing timber. When freshly emerged, in early autumn, the wide wing borders are yellow; after hibernation, in the spring, they are generally whitish. The caterpillars feed on willow, sallow or birch and are gregarious till nearly fully fed. Many attempts have been made to establish the Camberwell Beauty in England, but all have failed.

123 Comma, *Polygonia c-album* L. Expanse 1¾-2 inches.

Takes its name from the silvery C-shaped mark on the underside of the hindwing. Appears on the wing in early autumn, then hibernates and reappears in the spring. These spring butterflies give rise to a summer brood which flies in July. Caterpillar on hop, stinging nettle and currant. Restricted fifty years ago to the Wye Valley, it has since spread again through all the southern counties, most of the Midlands and even further north.

124 Queen of Spain Fritillary, *Issoria lathonia* L. Expanse 1¾ inches.

This lovely butterfly is unfortunately only an exceedingly rare casual visitor to England, coming almost certainly from southern Europe. It has been seen in Kent more often than in any other county. The caterpillar feeds on wild pansy and dog-violet.

125 Dark Green Fritillary, *Argynnis aglaia* L. Expanse 2¼-2½ inches.

The female of this handsome swift-flying butterfly is rather larger than the male, with more rounded, paler wings and heavier markings. Flies in July and early August throughout the British Isles in localities where violets, on which the caterpillars feed, grow freely.

126 Silver-washed Fritillary, *Argynnis paphia* L. Expanse 2½-3 inches.

Our largest British Fritillary. Flies in

July and August in most large woodland areas in the Midlands and southern England. Scarcer in the north and in Ireland. The female occurs, especially in Hampshire, in two forms, one much like the male in colour, the other (var. *valesina*) very much darker, almost olive-green. The caterpillar feeds in the spring on violets, after hibernating (without feeding at all) immediately after hatching from the egg.

127 Scotch Argus, *Erebia aethiops* Esper Expanse 1¾ inches.

Butterflies of the genus *Erebia* are characteristic of alpine regions. The Scotch Argus occurs in England only in the most northern counties, becoming commoner and more widespread in Scotland. It flies only when the sun is shining, in areas where the grasses grow, such as Tussock Grass, on which the caterpillars feed.

128 Grayling, *Satyrus semele* L. Expanse 2¼ inches.

Flies in July and August particularly on rather rough hillsides, where it is fond of settling on the ground with its closed wings pointing directly at the sun. Its under side harmonises with the background and makes it very difficult to see. Locally common throughout the British Isles. Caterpillars feed on a variety of coarse grasses.

129 Small Heath, *Coenonympha pamphilus* L. Expanse 1¼ inches.

Probably the commonest butterfly in the British Isles, occurring on grasslands everywhere, and on the wing from May to September. Caterpillar feeds on many kinds of grasses.

130 Wall Brown, *Pararge megaera* L. Expanse 1½-2 inches.

Well-named from its habit of basking on hedge-banks, walls and stones. Flies in May and again in August (sometimes even later) throughout the British Isles except northern Scotland. Cock's foot grass is the main food of the caterpillar.

131 Green Hairstreak, *Callophrys rubi* L. Expanse 1 inch.

One of the first spring butterflies to appear, early in May. Its flight and colour make it difficult to see on the wing. Common where it occurs as a rule, and present throughout the British Isles. Caterpillar feeds on the flowers, buds and seeds of gorse and broom.

132 Large Copper, *Lycaena dispar* Haworth Expanse 1½ inches.

This brilliant shining copper-coloured insect used to be common, more than one hundred years ago, in the fen country, where its caterpillar fed on the Giant Water Dock. Its disappearance was due less to the drainage of fens than to the cupidity of collectors.

133 Small Copper, *Lycaena phlaeas* L. Expanse 1 inch.

Generally to be found on the wing in most of the summer months, in two or three broods, depending on the season. The midsummer butterflies are duskier than those of spring and autumn. Throughout the British Isles. Caterpillar feeds on Sorrel and Dock.

134 Silver-studded Blue, *Plebejus argus* L. Expanse 1 inch.

Derives its name from the silvery scales in the dark spots near the margins of the underside of the hind wings. Local throughout the British Isles, but usually abundant where it occurs, which is chiefly on heathlands. Caterpillar feeds on the flowers of gorse and broom.

135 Chalk Hill Blue,
Lysandra coridon Poda
Expanse 1½ inches.

Found only, as its name implies, in chalk or limestone country where the Horseshoe Vetch grows. Caterpillar feeds exclusively on this vetch. The female is usually entirely brown on the upper side, except for a few orange spots near the hind wing margin, but there are varieties bearing a variable amount of blue scaling on the upper side.

136 Death's Head Hawk Moth,
Acherontia atropos L.
Expanse 4½ inches.

The largest British Moth, but only doubtfully native. Migrants from the south reach southern England during the summer months in most years and lay eggs on the Potato haulms. When the potatoes are lifted the pupae are turned up by the diggers; it seems unlikely that the moth survives the English winter. Another name for the Death's Head is Bee Robber, due to the fact that it sometimes raids beehives in search of the honey. Both the moth and its caterpillar are capable of making an audible squeaking noise. The moth derives its name from the markings on the thorax.

137 Eyed Hawk Moth,
Smerinthus ocellatus L.
Expanse 3½ inches.

Flies, at night, during May and June. Found throughout the British Isles, except norhern Scotland. By day rests with the eye spots of the hind wing covered by the forewings. The long proboscis of the moth enables it to reach the nectar of tubular flowers, and in doing so it incidentally helps to pollinate them. The caterpillar feeds on the leaves of willow and sallow and sometimes apple trees.

138 Privet Hawk Moth,
Sphinx ligustri L.
Expanse 4½ inches.

In Britain it is rarely found outside the southern counties of England where it is not uncommon in June and July. The moth is much more rarely seen, however, than the very handsome caterpillar, which feeds chiefly on privet, but also on lilac and sometimes ash.

139 Pine Hawk Moth,
Hyloicus pinastri L.
Expanse 2 inches.

Until recently only to be found in Britain in certain pine plantations in East Anglia. Now present in most of the southern counties of England, its spread probably helped by the extensive planting of coniferous trees in those counties in recent years. The moth flies during June and July resting by day on pine trunks. Caterpillar feeds on the needles of pine.

140 Bedstraw Hawk Moth,
Celerio galii v. Rott.
Expanse 2¾-3½ inches.

A rare migrant from southern Europe reaching the British Isles occasionally in May but more often in late summer. The conspicuous caterpillar sometimes found wild feeding on bedstraw and though generally green, may be brown and even black.

141 Elephant Hawk Moth,
Deilephila elpenor L.
Expanse 2½ inches.

Found throughout England and Wales

Ireland and southern Scotland in June. Visits flowers by night. Caterpillar feeds at night on willow-herb and bedstraw, especially in the neighbourhood of water; when disturbed it draws back its head, which causes the eye-like markings on its side to swell out, producing a snake-like appearance, though it is quite harmless.

142 Puss Moth, *Cerura vinula* L.
Expanse 2½-3 inches.

The night-flying moth rests by day mainly on tree trunks. It is less often seen than its handsome caterpillar, which feeds on poplars and willows. When disturbed, the caterpillar raises its head and draws it back, at the same time whipping its forked tail from side to side. The whole sudden action has a decidedly frightening effect.

143 Buff Tip Moth,
Phalera bucephala L.
Expanse 2½ inches.

Occurs in June and July throughout the British Isles. When at rest, with the wings furled along its back it resembles a piece of broken twig. The caterpillars, which are gregarious until nearly full grown, feed on the foliage of almost any tree.

144 Vapourer Moth,
Orgyia antiqua L.
Expanse 1¼ inches.

Particularly common in urban districts in the south, where its caterpillar teeds on a great variety of ornamental and shade trees. Less common in Ireland and Scotland. Flies during July to September. Female wingless. The tiny newly hatched caterpillars are sometimes distributed by the wind.

145 Small Eggar,
Eriogaster lanestris L.
Expanse 1½-1¾ inches.

Fairly widespread in southern England, rarer in the north of Scotland. Flies in February and March. Female larger and stouter than the male. Caterpillars live gregariously on a web of silk spun over a branch of the hawthorn or sloe on which they feed.

146 Black Arches,
Lymantria monacha L.
Expanse 1½-2 inches.

Occurs in most woodland areas from Yorkshire southwards, in July and August. Caterpillar feeds on foliage of oak and other trees, in the bark of which it makes a cocoon before turning into a chrysalis. Though not troublesome in the British Isles, the Black Arches is a serious pest of forest trees in some continental countries.

147 Fox Moth, *Macrothylacia rubi* L.
Expanse 2¼-2½ inches.

Throughout British Isles. The male flies wildly by day, seeking the female, in May and June, especially near open heathland. The larger female flies at dusk and evening. Caterpillar feeds on bramble, heath etc. and its hairs are irritating to the skin.

148 Oak Eggar,
Lasiocampa quercus L.
Expanse 2½-3 inches.

Common throughout the British Isles. In Scotland, Ireland and northern England it is darker than in the south and this variety is known as *callunae*. The moth flies in June, July or August, according to locality, and the caterpillar, which sometimes may take one or two years to reach full growth, feeds on hawthorn, blackthorn, heather and ivy.

149 Emperor Moth,
Saturnia pavonia L.
Expanse 2½-3 inches.

Throughout the British Isles, flying in

April and May, the male by day, the much larger and paler female at dusk. Prefers heath and marshland. Caterpillar feeds on great variety of bushes and shrubs, such as heather, bramble, sallow and sloe; its tough cocoon is furnished at one end with stout fibres arranged in such a way that, though the moth can emerge easily, enemies cannot enter.

150 Heart and Dart,
Agrotis exclamationis L.
Expanse 1½ inches.

Very common except in northern Scotland. Flies in June and July. Caterpillar rests in soil and feeds on many low-growing plants such as grass, chickweed, plantain and turnip, sometimes being quite a pest, at night.

151 Setaceous Hebrew Character,
Amathes c-nigrum L.
Expanse 1½ inches.

Abundant everywhere in the British Isles. Double brooded, flying in May and again in September, except in the north where it flies in June and July. Caterpillar feeds on dock, plantain, dandelion, chickweed etc. and hides in the soil by day.

152 Great Brocade, *Eurois occulta* L.
Expanse 2½ inches.

Found regularly in Scotland. Occasionally in England, even in the south, probably as migrants, in August. Caterpillar feeds in autumn on dandelion, knot grass etc. and in the spring after hibernation on low-growing sallow, bramble etc.

153 Large Yellow Underwing,
Triphaena pronuba L.
Expanse 2¼ inches.

Common everywhere, often abundant. The main brood in June and July, often a partial second brood in August and September. Caterpillar feeds on almost any low-growing plant and is often a pest in flower and vegetable gardens.

154 Broom Moth, *Ceramica pisi* L.
Expanse 1½ inches.

The colour of the forewings is a warm reddish brown. Flies in June and July throughout the British Isles. Caterpillar feeds mainly on broom, also on bramble, bracken, sallow etc. on which it can often be found by day.

155 Feathered Gothic,
Tholera popularis F.
Expanse 1¼-1¾ inches.

Flies in August and September throughout England, Wales and Ireland; in southern Scotland local. The caterpillar feeds at night on matgrass and other similar grasses.

156 The Claddagh, *Luceria virens* L.
Expanse 1½ inches.

This lovely moth was not known to exist in the British Isles until it was found in 1949 in the Burren of Clare in western Ireland. The caterpillar feeds in and around the roots of various grasses and the moth flies in July.

157 The Shark, *Cucullia umbratica* L.
Expanse 2¾ inches.

Flies by night in June and July. Rests by day on tree trunks, palings etc. where it is very difficult to see. Caterpillar feeds during July to September on sow thistle and allied plants.

158 The Sallow, *Cirrhia icteritia*
Hufnagel Expanse 1½-1¾ inches

Flies in September. Very variable in its markings which may be much reduced in extent. Caterpillar feeds at first in sallow catkins; afterwards on many low-growing plants. Common except in northern Scotland.

159 Bird's Wing, *Dypterygia scabriuscula* L. Expanse 1½ inches.

The popular name is based on the shape of the pale markings on the forewing. Common in southern England, rare or absent northwards. Flies in May and June, and the caterpillar feeds on dock, sorrel, knotweed and allied plants.

160 Clouded Border Brindle, *Apamea crenata* Hufnagel Expanse 1½-$1^{9}/_{10}$ inches.

Common everywhere and often abundant. Flies in June and July and rather variable. Caterpillar feeds from August to April on almost any kind of low-growing plant.

161 Copper Underwing, *Amphipyra pyramidea* L. Expanse 2-2¼ inches.

Flies from late July to September in the larger woodlands of southern England and Ireland, becoming rare or absent northwards. Caterpillar feeds in the spring on the foliage of most woodland trees.

162 The Dunbar, *Cosmia trapezina* L. Expanse 1¼ inches.

On the wing in July and August. Common from southern Scotland southwards; rare to the north. Caterpillar feeds on most woodland trees and is also strongly cannibalistic.

163 The Silver Y., *Plusia gamma* L. Expanse 1½ inches.

A well known migrant. Arrives sometimes in more than one wave during the summer. Commonest in late summer and autumn. Caterpillar feeds on almost any kind of low-growing plant.

164 Burnished Brass, *Plusia chrysitis* L. Expanse 1½ inches.

Occurs throughout the British Isles in July and August, fairly commonly. Caterpillar feeds on nettles and pupates in a cocoon on the leaves.

165 Clifden Nonpareil, *Catocala fraxini* L. Expanse 3½ inches.

Until recently a very rare migrant. Appears now to be established in one small area in Kent where it flies in July. Caterpillar feeds on aspen and poplar.

166 Red Underwing, *Catocala nupta* L. Expanse 3 inches.

Common in south and east England but rare or absent elsewhere. Often to be seen at rest on tree trunks with fore wings covering the hind wings. Caterpillar feeds on poplar and willow.

167 Orange Underwing, *Brephos parthenias* L. Expanse 1¼ inches.

Moth flies high during sunshine in March and April round birch trees, on which the caterpillars feed, firstly on the catkins, then the leaves. Northwards as far as southern Scotland.

168 Large Emerald, *Hipparchus papilionarius* L. Expanse 1¾-2 inches.

Flies in July through Britain, except northern Scotland, in woods and open country where birch, beech and hazel, on which the caterpillars feed, grow freely. Caterpillar hibernates when very small, on exposed twigs, and feeds up in the spring.

169 Blood Vein, *Calothysanis amata* L. Expanse 1¼ inches.

The thin diagonal red line, exposed

fully even when the insect is at rest, gives this moth its popular name. Common in southern counties, thence rarer northwards, in June and July. Caterpillar feeds on dock, knotgrass etc.

170 Purple-barred Yellow,
Lythria purpuraria L.
Expanse $\frac{3}{4}$ inch.

Only doubtfully British, two specimens having been taken once near Perth in 1861. Caterpillar said to feed on oak, blackthorn and sorrel.

171 Shaded Broad-bar,
Ortholitha chenopodiata L.
Expanse $1\frac{1}{4}$ inches.

Common almost everywhere in July and August in rather open country: at one time known as the 'Aurelian's Plague'. Caterpillar feeds on clover, vetch, grasses etc.

172 Clouded Border,
Lomaspilis marginata L.
Expanse $^{7}/_{8}$ inches.

Generally common throughout Britain, except the extreme north, especially in moist places where sallows abound, principally in May and June. Caterpillar also on willow and aspen.

173 Winter Moth,
Operophtera brumata L.
1 inch. (male).

The male flies in winter, mainly in December. The female is wingless, but after emerging from the chrysalis in the soil, climbs the trees on which she lays her eggs. It is to prevent this that greasebands are placed around the stems of fruit trees. In some years the caterpillars are so abundant that much damage is done to the foliage of the many trees, including fruit trees, on which they feed.

174 Magpie or Currant Moth,
Abraxas grossulariata L.
Expanse $1\frac{1}{2}$ inches.

Both caterpillars and moth are conspicuously coloured, warning predators that they are not very tasty. The moth occurs throughout Britain and is very variable. Caterpillar feeds on currant, blackthorn, gooseberry, euonymus etc.

175 Orange Moth, *Angerona prunaria* L. Expanse $1^{5}/_{8}$ inches.

This extremely variable moth flies in July. Extends northwards as far as Yorkshire, but not commonly. The caterpillar feeds on blackthorn, privet, plum etc., pupating between leaves in June.

176 Mottled Umber,
Erannis defoliaria L.
Expanse $1\frac{1}{2}$ inches.

Male moth flies in October to December. The pattern of wing markings is very variable. Female is wingless. The brightly coloured looper caterpillar feeds in the spring on oak and birch, and many other trees and is often so abundant as to strip the young leaves completely.

177 Bordered White,
Bupalus piniarius L.
Expanse $1^{3}/_{8}$ inches.

Caterpillar feeds from August to October on pine needles and, with the great increase in the planting of coniferous trees, has already become a pest in some areas. The moth flies in May and June throughout Britain.

178 Garden Tiger, *Arctia caia* L.
Expanse $2\frac{1}{2}$-3 inches.

The very active furry caterpillar of the Tiger Moth is more familiar than the moth. Known as the Woolly Bear, it is freqeuntly found feeding on a great variety of low-growing plants, especial-y in the spring after hibernation. The moth flies in July and August, and is generally common.

179 Clouded Buff, *Diacrisia sannio* L.
Expanse 1½ inches.

Inhabits open country such as heaths and mosses, throughout Britain. Male will sometimes fly by day, but the female seldom before dusk, in June and July. Caterpillar feeds on dandelion, dock, plantain and other low-growing plants.

180 White Ermine,
Spilosoma lubricipeda L.
Expanse 1½ inches.

Common throughout Britain, the moth is frequently attracted to light at night, in June and July. Very active caterpillar suggests a small, pale Woolly Bear (see Tiger Moth) and similarly feeds on all kinds of low-growing plants.

181 Hornet Moth, *Sesia apiformis* L.
Expanse 1½ inches.

Very aptly named, for in flight this clear-wing moth may easily be mistaken for a wasp or hornet. It flies in May and June in southern and eastern England, rarely as far north as Scotland, and in Ireland. The caterpillar feeds in the roots and stems of poplar and takes two years to reach full growth.

182 Ghost Swift, *Hepialus humuli* L.
Expanse 2½-3 inches.

Flies at dusk in grassy places, the male with a curious swinging flight, like a pendulum, to and fro, sideways; the female much less easily seen, in straight level flight. Caterpillar lives in the soil feeding on the roots of coarse grasses and other vegetation. The chrysalis makes its way to the surface of the soil when the moth is due to emerge. Common in all suitable localities.

183 Six-spot Burnet,
Zygaena filipendulae L.
Expanse 1½ inches.

Widely distributed on chalk downs and rough hillsides where the trefoils, clover and similar plants grow on which the caterpillar feeds. Day-flying, conspicuous and rather sluggish. The silvery cocoons are conspicuous on grass stems.

184 Goat Moth, *Cossus cossus* L.
Expanse 3-3½ inches.

Occurs throughout Britain, accept northern Scotland, the moth flying in June. Derives its popular name from its caterpillar, which has a very goat-like smell and tunnels the wood of elm, ash, willow, apple and even oak, taking three years to reach full growth.

185 Wax Moth, *Galleria mellonella* L.
Expanse 1¼-1½ inches.

Moth flies from July to October, locally in England and Ireland, at night. Caterpillar feeds on old honeycomb, in which it forms stout silk-lined burrows and later a tough cocoon.

186 Mill Moth or Mediterranean Flour Moth, *Ephestia kuehniella*
Zeller Expanse 1 inch.

Common in flour mills, granaries, bakeries etc. throughout Britain during most of the summer months. Caterpillars feed on flour, which they cover with silken webs, making the handling of the flour very difficult and otherwise spoiling it.

187 Green Oak Beauty,
Tortrix viridana L.
Expanse ¾-1 inch.

Abundant in Britain at least to southern Scotland. Moth flies in June and July. In spring the caterpillars are sometimes so numerous as almost to strip the oaks. When full fed they descend to the ground on silken threads to pupate.

188 Codling Moth, *Cydia pomonella* L. Expanse ½ inches.

An important pest of apples, occurring wherever apples are grown. Moth flies at dusk in May and June and the female lays her eggs on the young fruits, into which the caterpillar, sometimes known as the Apple Worm, immediately bores.

189 Pea Tortrix, *Laspeyresia nigricana* Stephens Expanse ½ inch.

Fairly common from southern Scotland southwards. Moth flies in June and July. Eggs are laid on or near the young pods of peas in which the caterpillar feeds.

190 Pine Tortrix, *Evetria buoliana* Schiffermuller Expanse ½ inch.

Moth flies in July and August in pine woods but is commoner in the south than in Scotland. It is one of several very similar species the caterpillars of which bore in the young shoots of various pine trees, doing considerable damage.

191 Small Ermine, *Hyponomeuta evonymella* L. Expanse ¾ inch.

The caterpillars of the Small Ermines, of which there are several very closely related species, live in silken webs or tents which they spin over the young shoots and leaves of apple, plum, haw-thorn, euonymous and other trees, according to the species. They feed up in May and June and the moths fly in July and August in most English counties and in Ireland.

192 Apple Fruit Miner, *Argyresthia conjugella* Zeller Expanse ½ inch.

Normally the caterpillar feeds in berries of mountain ash, but sometimes mines the surface of apples. Moth occurs throughout Britain, in June and July.

193 Clothes Moth, *Tineola bisselliella* Hummel Expanse ½ inch.

Female usually noticeably larger than the male, but less often seen. Damage to clothes, carpets etc. is caused by the caterpillars which feed on all kinds of woollen and hairy material. Moth flies normally in June and July, but in centrally heated houses it may breed all the year round.

DIPTERA
Flies, Mosquitos, Midges

194 Winter Gnat, *Trichocera hiemalis* Degeer Expanse ½ inch.

Congregates in dancing swarms in winter, sometimes even when snow is about. Abundant everywhere. Larva in decaying vegetable matter.

195 Common Daddy Long Legs, *Tipula oleracea* L. Expanse 1½-2 inches.

Generally distributed in Britain in May and June, and often later. The very similar *Tipula paludosa* flies mainly in August and September. The larvae of both are known as Leather Jackets and often do considerable damage to lawns and pastures by feeding on the roots of grasses below the surface.

196 Spangle-winged Mosquito, *Anopheles maculipennis* Meigen Expanse ½ inch.

Occurs throughout Britain in a coastal variety, *atroparvus,* and an inland form, *masseae,* the former being the more important as a potential carrier of malaria. Only the females 'bite'. The very characteristic resting attitudes of the mosquito and of its larva are illustrated. Larvae in rather still, unshaded water.

197 Common Gnat, *Culex pipiens* L.
Expanse ½ inch.

Abundant throughout Britain. Larva breeds in almost any stagnant water, especially if slightly contaminated with rotting organic matter. Only the females survive the winter, to breed again in the spring, and they rarely, if ever, bite man.

198 Ringed Mosquito,
Theobaldia annulata Schrank
Expanse ½-¾ inch.

The largest British mosquito and common everywhere. Breeds in stagnant ditches fouled by sewage, as in the neighbourhood of farms. Only the female survives the winter, hibernating in cellars, sheds, hollow trees etc. Only the female bites (as in all mosquitoes), and her bite is more vicious and more often poisonous than that of any of the other species.

199 Hairy Moth Fly,
Psychoda alternata Say
Expanse ¼ inch.

The fly is often to be seen on windows, but is common out-of-doors in damp places. Larvae live in muddy places rich in decaying matter, such as sewage farms, waste pipes etc. Everywhere common.

200 River Fly or Buffalo Gnat,
Simulium reptans L.
Expanse ½ inch.

Male black, female greyish. The females bite man freely and are a great nuisance. Larvae live in running water, attached to stones by means of a kind of sucker at their hind ends. Widespread.

201 Harlequin Fly,
Chironomus plumosus L.
Expanse ¾ inch.

Larvae of this very abundant fly are the familiar blood worms of water butts, stagnant pools etc. When not swimming with their curious twisting movement they inhabit tubular mud shelters on the bottom.

202 Red-legged Bibio,
Bibio pomonae F. Expanse ¾ inches

On the wing in summer and again in autumn, chiefly in hilly country. Larva lives in soil. The commoner St. Mark's Fly, so-called because it usually emerges about St Mark's Day, differs in being entirely black and often flies in swarms.

203 Horse Fly or Gad Fly,
Tabanus bovinus L.
Expanse 1½ inches.

A woodland and forest insect, the male visits flowers in search of nectar, but the female pierces the skin and sucks the blood of cattle, and will also on occasion attack people.
Larva lives in moist soil and is predacious.

204 Clegg, *Haematopota pluvialis* L.
Expanse ¾ inches.

A vicious biter, which arrives quite silently and attacks man and beast. Common in southern England in wooded country from May to September. Larva lives in soil.

205 Warble Fly, *Hypoderma bovis* L.
Expanse 1 inch.

The fly lays its eggs on the hairs on the legs of cattle, whence they are licked into the mouth. The young larva penetrates the skin and by degrees migrates through the host's body to come to lie finally just below the skin of the back, where it gives rise to a swelling known as a warble. Sufficiently common to be of economic importance.

206 Common Horse Bot Fly,
Gasterophilus intestinalis de Geer
Expanse 1 inch.

The fly lays its eggs on the hair of the horse's legs whence they are apparently licked by the horse into the mouth. Entering by way of the lips, they eventually attach themselves to the stomach lining, and when full-fed release their hold and are passed out with the horse's droppings. They then pupate in the soil.

207 Assassin Fly,
Laphria marginata L.
Expanse 1 inch.

Does not seize its prey on the wing, but lies in wait and pounces on it, often capturing insects larger than itself which it appears at once to paralyse. Common in woodlands in southern England.

208 Large Bee Fly,
Bombylius major L. Expanse 1 inch.

When in flight greatly resembles a small hovering bumble bee. Appears in April and May, in southern England. Larva is a parasite feeding on the grubs of bees.

209 Swarming Hover Fly,
Scaeva pyrastri L. Expanse 1 inch.

Common in the British Isles, sometimes in swarms on the south coast, and generally distributed. Female often without yellow markings on body. The active spear-headed larvae are predacious on aphids and most useful in keeping them in check. Flies from May to October.

210 Drone Fly, *Eristalis tenax* L.
Expanse 1 inch.

Common in the British Isles. The larva lives in liquid putrefying matter and is known as the rat-tailed maggot, the long tail being a breathing tube which is pushed above the surface of the liquid. The fly is so like a bee that it is probably the 'bee' which the ancients thought was 'generated' from the rotting body of a lion.

211 Vinegar Fly,
Drosophila transversa Fallen
Expanse ½ inch.

The heavy slow flight is very characteristic of this species and its many close allies which frequent breweries and places where fermentation processes are carried on. Because they are very easily reared in captivity and breed rapidly they have proved an excellent subject for experimental research in genetics.

212 Flesh Fly, *Sarcophaga carnaria* L.
Expanse 1 inch.

Occurs throughout the British Isles between May and October. Female does not lay eggs, but young larvae, in decomposing or fresh flesh of almost any animal or in manure.

213 Green Bottle Fly,
Lucilia caesar L. Expanse ¾ inch.

Generally common throughout the summer months. Fly lays its eggs on carcasses (sometimes on butcher's meat if not properly protected) and in two or three days the larvae are fully grown.

214 Blow Fly, *Calliphora vomitaria* L.
Expanse 1 inch.

A very common and troublesome fly, especially in hot weather, since it lays its eggs not only in carrion (its natural habit) but also in any meat to which it can gain access. The resultant maggots ('gentles') feed up very rapidly and pupate within the hardened last larval skin.

215 Deer Bot Fly,
Cephenomyia auribarbis Meyer
Expanse 1 inch.

The life history of this fly is like that of the Sheep Bot fly *Oestrus ovis*. The viviparous females deposit young larvae in the nostrils of the deer. The larvae live in the nasal or throat passages, attached by their mouth hooks, feeding on secretions of the host. When full-fed they loosen their hold and are ejected, to pupate in the ground.

216 House Fly, *Musca domestica* L.
Expanse ½ inch.

Breeds in almost any kind of moist kitchen refuse, rubbish dumps, if moist, and particularly manure. It is a danger to health through transporting disease germs as it flies from filth to food, which it contaminates. Breeds very rapidly during the summer months, but in Britain seems only able to survive the winter in well-heated premises such as hotel kitchens and restaurants and boiler rooms.

217 Lesser House Fly,
Fannia canicularis L.
Expanse ½ inch.

Only very slightly smaller than the House Fly. The males have a very characteristic habit of flying round, 'patrolling', back and forth, rather slowly, around hanging lamps etc. indoors; the females, like the common House Fly, are more inquisitive in visiting foodstuffs. Breeds in rubbish heaps.

218 Forest Fly, *Hippobosca equina* L.
Expanse ¾ inch.

This rather flat parasitic fly is found chiefly in the New Forest where both sexes settle preferably on the more exposed and soft surfaces of forest ponies and cattle, and suck their blood. Their shape and claws enable them to move quickly, crab-wise, through the hairs of the host. The female is viviparous, giving birth to full grown larvae which at once pupate. On the wing from May to October.

219 Sheep Ked, *Melophagus ovinus* L.
Length ¼ inch.

Also known as the Sheep-tick or Sheep-louse, this completely wingless fly passes its whole existence in the fleece of sheep. It looks more like a spider than a fly, and, like the Forest Fly, is viviparous.

SIPHONAPTERA - Fleas

220 Dog Flea, *Ctenocephalides canis* Curtis Length 2-3 mm.

Usually darker than the human flea, almost black. Also distinguished by having a 'comb' of extra stout broad spines just behind the head. Its shape, flattened from the sides, and its strongly backwardly directed spines, enable it to slip through the hairs of its host and maintain a hold. The hind legs are specially developed for jumping. Rarely bites man, but occurs on cats as well as dogs.

221 Flea, *Pulex irritans* L.
Length 2-4 mm.

Lacks the 'comb' behind the head that distinguishes the Dog and Cat Fleas. Not nearly so common as it used to be. The grubs, like those of the Dog Fleas, are thin, maggot-like objects and live amongst organic refuse, such as accumulates in the cracks between floor boards and similar places.

HYMENOPTERA
Ants, Bees and Wasps

222 Honey Bee or Hive Bee, *Apis mellifica* L. Length Male 16 mm; Female 15-20 mm, Worker 10-13 mm.

There are many different strains or races of the Honey Bee, but none is native to Britain. The occasional 'wild' colonies that occur are escapes. Northern races are rather dark brown, those of Italy redder. The original home of the species was probably in the Near East, but now, owing to man's activities the hive bee is virtually cosmopolitan.

The male is readily distinguished by its very large eyes and 13-jointed antennae; the female or Queen has 12-jointed antennae, well separated eyes, and her long, rather pointed body projects beyond the wings when these are folded. The worker is the smallest of the three castes and has pollen baskets developed on its hind legs. It is the ubiquitous worker that is the most familiar caste. The drones, being males and therefore devoid of ovipositors, have no stings. A flourishing hive may contain 50,000 to 80,000 workers, one queen and a few hundred males. The queen, together with many of the workers survive the winter, living on the honey gathered in summer, or on sugar provided by the hive owner, but the drones are driven out on the approach of winter. The honeycomb is constructed of wax which is secreted by pores on the underside of the bee's body, and a resinous substance called propolis, collected from the buds of trees. The female lays a single egg in each brood cell, and the young larvae are at first fed on a special diet secreted by the workers, later on honey and digested pollen. The royal cells, in which queens develop, are larger than those of the workers and drones and rounded rather than hexagonal, and the larvae in them are fed on a special diet. The members of a colony are considered to be held together and to recognise each other through a secretion of the queen's, called queen substance, which pervades the whole nest and is shared by all its inhabitants. Lately it has been shown that the workers can communicate information to each other as to the distance and direction of sources of nectar by curious dances, which they perform within the hive.

223 Early Mining Bee, *Andrena albicans* Muller Length 8-11 mm.

Common in southern England and not very particular as to the kind of soil in which it sinks its shafts. From these shafts there branch off cells which the Bee stores with honey and pollen for the use of the grub that emerges from the single egg deposited in each. The bee appears in early spring, as soon as the sallow catkins open and the colt's foot flowers.

224 Patchwork Leaf-cutter Bee, *Megachile centuncularis* L. Male 8-10 mm, Female 11-12 mm.

The leaves chiefly cut out by this Bee are those of rose trees. The neatly cut pieces are used to make cylindrical brood cells, usually five or six, one above the other in burrows in earth or rotting wood. Most active in June and July.

225 Hill Cuckoo Bee, *Psithyrus rupestris* Fab. Length Male 15-18 mm, Female 18-25 mm.

The smoky wings and the absence of pollen-carrying apparatus on the hind legs distinguish this bee from the

Redtailed Bumble Bee (Plate 52 : 1) on which it is parasitic. It forces its way into the nest of the host, often killing the mother bee, but not the workers, whose services are needed to feed its offspring.

226 Buff-tailed Bumble Bee,
Bombus terrestris L.
Male 16-18 mm, Female 16-22 mm, Worker 10-16 mm.

The nest of this large and common Bumble Bee is started in the early spring by a female which has hibernated after the males and workers died off in the autumn. It usually is made in a hedge bank or the old nest of a shrew or field mouse, and consists of pieces of grasses and debris. In the centre the bee makes a kind of platform of pollen and honey on which she lays a batch of eggs. So soon as this first batch of workers has developed, the queen lays further batches and the workers begin to take over the nest, which grows gradually larger till the advent of colder weather brings activity to an end. The species is found throughout Britain and performs most valuable service to man as a pollinator, especially of lucerne and red clover.

227 Red-tailed Bumble Bee,
Bombus lapidarius L.
Male 14-16 mm, Female 20-21 mm, Worker 10-14 mm.

Appears on the wing about mid May, rather later than *Bombus terrestris*. Common generally though rare or absent in northern Scotland. Its life history is very similar, but it perhaps visits a wider variety of flowers.

228 Common Carder Bee,
Bombus agrorum Fabricius.
Male 13-15 mm, Female 10-14 mm, Worker 17-19 mm.

Common throughout the British Isles, nesting always on or above the ground, even in old pots and pans. The life history closely resembles that of other species of *Bombus*.

229 Big-headed Digger Wasp,
Ectemnius cavifrons Thomson.
Expanse 1 inch.

Flies in June, July and August throughout Britain. Makes its cells singly in rotten wood and provisions them with flies, especially blue and green-bottle flies.

230 Wall Mason Wasp,
Odynerus parietum L.
Expanse 1 inch.

Makes a nest of mud cells in cracks in walls, window frames, loose brickwork etc, often as many as ten or more at a time. Provisions the cells with paralysed caterpillars for its grubs to feed on. Common in most parts of the country.

231 Hairy Sand Wasp,
Podalomia viatica L.
Length 15-22 mm.

A robust handsome species with a furry appearance, burrowing in sandy places. Provisions its nest with large caterpillars (sometimes larger than itself) of the kind that hide by day in or near the surface of the soil, such as cutworms. A southern species, on the wing in spring and early summer.

232 Red-banded Sand Wasp,
Sphex sabulosa L.
Length 12-20 mm.

Like the previous species this wasp provides for its offspring a larder composed of large caterpillars; however, instead of first finding the caterpillar, then burying it, it does its excavation first then drags its prey to it. Only one egg is laid in each shaft dug. Not uncommon in sandy places, even by the sea above high water mark.

233 Red-banded Spider Wasp, *Anoplius fuscus* L.
Length 10-15 mm.

A common species in sandy places, from April to August. It hunts spiders, which it stings and paralyses. It then drags them to suitable spots in which to sink shafts for their burial. When satisfied, the wasp lays an egg on the still living spider and fills in the shaft above it with sand and pebbles. Finally it covers up the disturbed area with miscellanous debris.

234 Common Wasp, *Vespa vulgaris* L.
Length 11-20 mm.

Like many bees, wasps have evolved a caste system of males, females and workers. Of these the females are much the largest, and alone survive the winter to carry on the species. Every queen wasp killed in the spring therefore means one less wasps' nest. Males can be recognised by their longer antennae. The nest is usually made underground and is roughly spherical, is built of wood pulp, and is suspended very often from the root of a tree. It consists of an outer covering, often of several layers, enclosing several horizontal combs of hexagonal cells which open downwards. In these cells the brood is reared, the cells being closed below when the grub is full-fed. The process of building up the nest and its population from the first few workers raised by the queen, is a slow one, which acounts for the fact that wasps are seldom numerous until late summer or autumn. There are several common wasps of the genus *Vespa* in Britain so much alike as to be confusing even to the expert. All are attracted to sweets, though in the main they are predacious on other insects. Only the females and workers can 'sting'.

235 Hornet, *Vespa crabro* L.
Length 20-35 mm.

This large wasp can at once be recognised by its size and dull reddish-brown colouring. It usually makes its nest in hollow trees. Though its sting is very painful, it is generally less aggressive than the common wasps. It occurs in most southern and midland counties and often flies at night, coming to street lamps and the lights of houses.

236 Heath Potter Wasp, *Eumenes coarctata* L.
Length 11-14 mm.

The popular name is based on this little wasp's habit of building round, flask-like mud cells for her offspring. Usually a stiff twig near the ground is chosen, sometimes stones or palings. Each cell is provisioned with small caterpillars which have first been stung, and then a single egg is suspended inside before the cell is closed. Fairly common in southern England, especially on sandy heaths.

237 Wood Ant, *Formica rufa* L.
Length 6-11 mm.

The largest British ant, and to be found in most woodlands throughout the country, especially in pine woods, where its large nests of pine needles and other fragments are often conspicuous, rising sometimes to a height of several feet. The main nest, however, consists of galleries and chambers below ground. It is a pugnacious creature, and when disturbed will not only bite but also discharge a droplet of stinging formic acid. Each nest contains several queens besides the males and innumerable workers. Swarming takes place in June, and mating occurs on the ground, not in the air. The wood ant is omnivorous, capturing insects, feeding on nectar and honeydew (the sweet excretion of aphids) and also the seeds of many plants. Many other insects, mites and other creatures share their nest with the ants, especially beetles. Some are merely tolerated; others are tended by the ants in return for their secretions which the ants eagerly seek.

238 Jet Ant, *Lasius fuliginosus* Latreille
Length Male and Worker 4-6 mm, Female 6-8 mm.

Nests principally in old trees, or among their roots, several nests often being connected by underground tunnels. Has a noticeable and not unpleasant odour, and makes well-defined tracks to the sources of its food supplies, which are mainly the excretion of aphids, insects and seeds. It frequently raids the nests of other ants and carries off their pupae. Occurs northward as far as Yorks and Lancashire.

239 Small Black Ant, *Lasius niger* L.
Length Male and Worker 3-5 mm, Female 8-9 mm.

Abundant throughout Britain and also known as the Garden Ant, owing to its liking for nesting under the stones of rockeries, paths etc. Frequently invades houses in search of sweet substances such as sugar and jam. Swarms in July and August, and when this occurs in towns and cities, as often happens, generally causes quite unnecessary consternation. Omnivorous in habit, feeding on seeds, nectar, insects and the secretions of aphids both harboured in the nest and free-living on plants near the nest.

240 Yellow Ant, *Lasius flavus* F.
Length Male 3-4 mm, Female 7-9 mm, Worker 3-5 mm.

Occurs throughout Britain, abundantly, nesting chiefly in fields, where it raises the earthy mounds often mistaken for mole-hills. Appears to live almost exclusively on the excretions of aphids and scale insects which inhabit its nest, but it also 'milks' the caterpillars of the Chalk Hill Blue Butterfly.

241 Pharaoh's Ant, *Monomorium pharaonis* L.
Length 2-2.5 mm.

A cosmopolitan ant, in England confined entirely to houses, especially well-heated buildings, where it can become a serious pest. Feeds voraciously on sweets, cake, also meat, butter and fats. A marriage flight (winged swarm) has not been seen in England.

242 Red Ant, *Myrmica rubra* L.
Length 3-6 mm.

Nests under stones or forms small thyme-covered nests on the surface, or even in tree stumps. Carnivorous, attacking other ants and insects; also has a fondness for sweet substances such as the honeydew of aphids, some of which live in its nests. Several different kinds of ants are probably confused under the name of 'Red Ant'. One of them carries off the young caterpillars of the rare Large Blue Butterfly and tends them in its nest.

243 Ruby-tailed Wasp, *Chrysis ignita* L. Length 6-9 mm.

This dazzlingly beautiful little wasp can sometimes be seen sunning itself on hot walls and stones. Its beauty belies its habits, for it parasitises the wall Mason Wasp (*Odynerus parietum,* Fig. 230), its grub consuming the food store laid up by the Mason Wasp for its own offspring.

244 Cherry Gall, *Cynips quercusfolii* L. Expanse ½ inch.

Causes round galls that appear on the ribs on the underside of oak leaves in the autumn, each containing one larva of the gall-wasp. In late autumn there emerge from these galls females only, which lay eggs in the resting buds of the oak. The larvae hatching from these give rise to violet 'egg' galls in spring, which are only two to three millimetres long and thus very small when compared with the cherry galls of the alternate generation. From these insignificant galls both males and females emerge about the end of May, the females laying eggs on the oak leaves to produce the Cherry Galls once again.

245 Robin's Pin Cushion, *Rhodites rosae,* L. Expanse 1/8 inch.

The characteristic galls made by this gall wasp are familiar on rose bushes, particularly wild roses. The wasp has no popular name, being known, as is usually the case, by the name applied to the gall its larvae produce. Known also as the Bedeguar. There are several, sometimes many larvae, in each gall: and there is no alternate generation like that of the Cherry Gall.

246 Horn-tail Ichneumon, *Rhyssa persuasoria* L. Expanse 1¼-1¾ inches.

The largest British 'Ichneumon Fly'; it is actually a parasitic wasp. When in action, with antennae waving in front and ovipositor stretched out behind, it measures fully 3½ inches. With its ovipositor it drills through the bark and solid wood of pines to lay an egg in the larva of the Giant Wood Wasp, *Sirex gigas,* (Fig. 249). How it locates this larvae is not known. Local, but widely distributed in pine woods.

247 Yellow Ophion, *Ophion luteus* L. Expanse 1-1½ inches.

Parasitises the caterpillars of the larger moths. Flies at dusk and later, often invading houses. Common in most areas.

248 Cabbage White Wasp, *Apanteles glomeratus* L. Expanse 5 mm.

The tiny yellowish silken cocoons often found surrounding a dead Cabbage White caterpillar are made by the grubs of this very small wasp after leaving the 'host' in which they have fed. The wasp is an important natural control on the butterfly.

249 Giant Wood Wasp, *Sirex gigas* L. Expanse Male 1½ inches. Female 2¼ inches.

Often mistaken for a hornet, but not a true wasp, as the absence of a waist indicates, and devoid of a 'sting'. The female bores in the wood of unsound pine trees with her ovipositor, and there lays an egg from which emerges a grub that tunnels in the wood for several years before becoming fully grown. Found throughout the country in pine woods, but never common.

250 Steel-blue Wood Wasp, *Sirex juvencus* L. Expanse 1¼-1½ inches.

Life history similar to that of the Giant Wood Wasp. In pine woods in England and Scotland. Not a native species, but now established in Britain. Often introduced with timber.

251 Willow Sawfly, *Pteronidea salicis* L. Expanse 2/3 inches.

One of the many kinds of sawflies that attack the foliage of willows. The larvae, which are like caterpillars but have more legs, skeletonise the leaves. Sawflies are so named because the female's ovipositor has a saw-like edge with which she cuts plant tissues when egg-laying. Common.

252 Yellow Cimbex, *Cimbex lutea* L.
Expanse 1½-2 inches.

The caterpillar-like larva is to be found on sallow and poplar throughout England and Scotland from May to July.

Protective Resemblance - Mimicry

253-261 Examples of Protective Resemblance

Figures 253 to 257 show examples of insects which escape detection when at rest by resembling and 'melting' into their background. In figures 258 to 261a Clearwing Moth and a two-winged Hover Fly are shown which look deceptively like the Common Wasp and so gain protection from their enemies by 'flying under false colours'. Most of these are described elsewhere, and cross-references are as follows:
253 see No. 71
255 see No. 128
258 see No. 181
259 see No. 209
260 see No. 234

Migrants and Casual Visitors

In the British Isles no insect is protected by law, as is the case in some continental countries. However, as certain rare and local species appeared to be threatened with extinction for one cause or another, the Royal Entomological Society of London drew up and published in 1951 a list ot Butterflies and Moths which it asked entomologists to refrain from catching and to protect in whatever way they could. The appeal met with a very satisfactory response. Only one of the species listed is illustrated in this book, namely the Swallow-tail Butterfly *(Papilio machaon)*. One other, Blair's Wainscot, has already become extinct it is feared, for its only known habitat was largely destroyed when it was greatly in demand by collectors—a combination of circumstances to which the loss of many insects that formerly inhabitated these islands can certainly be attributed.

262 New Zealand Stick Insect, *Acanthoxyla prasina* Westwood
Length 3½ inches.

This prickly stick insect was first found in England at Paignton, South Devon, in 1908. It is a native of New Zealand and it is considered almost certainly to have been imported to Paignton and the Scilly Isles (where it is also found) with growing plants brought from New Zealand in 1907. It should be looked for on such plants as roses, brambles and raspberries. As it seldom moves by day it is best sought with a lamp at night.

263 Red-veined Dragonfly, *Sympetrum fonscolombii* Selys
Expanse 2¼-2½ inches.

A very rare visitor to the British Isles which sometimes suceeds in establishing itself for two or three years. It generally arrives in July, if at all, and is strongly migratory in habit, ranging over the whole of Africa, Europe and much of Asia. The adults affect rather boggy ponds, and in such places the nymphs have been known very rarely to breed.

264 Banded Hover Fly, *Volucella zonaria* Poda
Expanse 1¾ inches.

This large and handsome Hover Fly might well be mistaken on the wing for a bee or wasp. Formerly a great rarity, in recent years it has become established in a number of southern counties, from Essex to Gloucestershire. The larva lives as a scavenger in the nests of bees and wasps.

265 Bath White, *Pontia daplidice* L.
Expanse 2 inches.

The Bath White is a native of the Mediterranean countries and the Near East, in some of which it is on the wing all the year round. It is rather a wanderer, but very rarely reaches the British Isles unless climatic conditions are particularly favourable. A second generation has been known to result from early migrants, as there is no lack in this country of the various cresses on which the caterpillars feed.

266 Milkweed Butterfly,
Danaus plexippus L.
Expanse 4 inches.

How this handsome butterfly, which is a native of North America, reaches the British Isles is still rather a mystery. It is certainly a great migrant, having at one time spread throughout the Pacific Islands. But the crossing of the Atlantic would seem too heavy a task even for such a strong flyer. It is now thought that it generally makes the passage accidently with the cargo in holds of ships, as most of the captures in England are concentrated near the ports. The caterpillar feeds on various milkweeds, none of which is a native British plant.

267 Colorado Beetle
Leptinotarsa decemlineata L.
Length 10 mm.

This serious and voracious pest of potato crops reached Europe from America and has now spread throughout most of western Europe. It is very easily recognised, fortunately, both as larva and adult. It passes the winter so deep in the soil that ordinary cultivation does not disturb it. In a bad outbreak the whole of the haulms may be stripped and eaten. Occasionally reaches Britain, probably through accidental importation, but possibly carried by strong winds (it is a strong flier) from the continent. Constant watch should be kept on potato fields, especially in southern England, and the discovery of the beetle or its grub at once reported to the Police.

SHORT BIBLIOGRAPHY

Since 1634, when the first work on British insects, *Insectorum sive minimorum Animalium Theatrum* by Thomas Moffett, was published, a great many others, ranging from the highly specialised scientific treatise to the purely popular, have made their appearance. Great additions to knowledge have also been made known through the publications of scientific societies. The volume of information now-a-days available is therefore extensive, yet it by no means covers the whole field. Though a great deal has been written of such popular insects as butterflies and moths, little is still known of the natural history of many of the more obscure kinds such as the smaller flies, the parasitic wasps and certain groups of beetles. The reader who wishes to learn more about British insects may find the works listed below useful.

Clegg, John. Freshwater Life of the British Isles. London, 1959.
Colyer, C.N. & Hammond C.O. Flies of the British Isles. London, 1951.
Daglish, E.F. Name that Insect. London, 1952.
Hickin, N.E. Caddis. London, 1952.
Hirons, M.J.D. Insect Life of Farm and Garden. Blandford, 1966.
Imms, A.D. Outlines of Entomology. London, 1947.
Imms, A.D. Insect Natural History. London, 1950.
Jou, Norman. A Practical Handbook of British Beetles, 2 vols. London, 1932.
Joy, Norman. British Beetles, their homes and habits. London, 1933.
Longfield, Cynthia. Dragonflies of the British Isles. London, 1949.
Miall, L.C. Aquatic Insects. London, 1903.
Ragge, D.R. Grasshoppers, Crickets and Cockroaches of the British Isles. London, 1965.
Rye, E.C. British Beetles. An introduction. London, 1890.
Sandars, E. An Insect Book for the Pocket. London, 1946.
Smart, J. Instructions for Collectors. London, 1949.
South R. Butterflies of the British Isles. London, 1924.
South, R. Moths of the British Isles. 2 vols. London, 1939.
Southwood, T.R.E. and Leston, D. Land and Water Bugs of the British Isles. London, 1959.
Step, E. Bees, Wasps, Ants etc. of the British Isles. London, 1932.
Stokoe, W.J. Butterflies and Moths of the Wayside and Woodland. London, 1939.

A series of more specialised *Handbooks for the Identification of British Insects* is in course of preparation by the Royal Entomological Society of London, but is not yet complete. Other books, in considerable variety, are available at most good booksellers.

INDEX

References are to numbered illustrations (ana associated numbered descriptions) and NOT *to pages.*

RACE A

I hardly dared look at Monkey as we dashed into the stable. My once beautiful horse was lying wretchedly on the ground, his sides contorting with his laboured breathing. I squeezed down next to him as the vet rose to fill his syringe, and I gently nursed his head and put my face on his muzzle.

'Oh, Monkey! I'm so glad I'm with you now.' I buried my face in his familiar smell and somehow, from deep inside him, he summoned the strength to move his ears.

Tears were welling in my eyes as I turned to the vet. He was tearing open the swab. 'You're not a moment too soon, Becky!'

Also by Ginny Elliot

Winning!

GINNY ELLIOT

RACE AGAINST TIME

Collins

An imprint of HarperCollins*Publishers*

A very special thanks to Fiona Holgate for her support and inspiration during the writing of this book.

And, in acknowledgement, and grateful thanks for the information supplied on the factual side of the book: Don Attenburrow, Lord Patrick Beresford, Dr Richard Manchee, Hana Mottlová, The Chelsea Football Club, Danilo Mosca and Linda Spendley.

First published by Collins in 1996
Collins is an imprint of HarperCollins*Publishers* Ltd,
77-85 Fulham Palace Road, Hammersmith, London W6 8JB

1 3 5 7 9 8 6 4 2

ISBN 0 00 675163 6

Printed and bound in Great Britain
by Caledonian International Book Manufacturing Ltd.,
Glasgow, G64

For all the wonderful horses I have ridden,
including the very special Dubonnet,
Priceless, Night Cap, Murphy Himself,
Welton Houdini, Master Crafts Man, and
for their minders, the girl grooms who have
helped care for them
over the years.

CHAPTER ONE

Red Rag was lying on the stable floor. He wasn't moving – and the only sound I could hear was the noise of his laboured breathing. Roughly, I elbowed my way through the semi-circle of people who were standing quietly watching the vet examine the sick horse.

'Let me through! Please! I've got the serum!' I flung myself down in the straw and shoved the package into the vet's hand. 'Oh, Monkey! Please don't die!' Tears were pouring down my cheeks and I cradled his head in my lap. 'Hang on for just another few hours! You can make it! Please try!'

The sick horse opened his eyes, moving his head a fraction, barely aware that I was there. He looked awful. His coat was matted and dull and his ribs shuddered as he struggled to breathe.

The vet quickly gave the injection and shook his head. Holding my hand, he said, 'Becky, I don't think he's going to make it. You'll have to prepare yourself for the worst!'

'No! No!' I shot up in bed, hot and clammy, as I slowly emerged from my blurred, half-world of dreams. It was a nightmare! The same one I'd been having for six months, ever since the deadly disease had swept through the trainers' yards. Awake now, tears still rolled down my cheeks as I remembered the dreadful chain of events that had changed my life. It had all started last summer.

Paddy, the new horse, was due to arrive. The alarm clock hadn't worked and Ned, my grandfather, had woken me up.

'Oh, no! Not today of all days!' Late, and in a blind panic, I leapt out of bed, grabbing whatever clothing lay to hand. Running into the bathroom, I made a face at the mirror. 'Sorry, hair! You'll just have to wait!'

I'd meant to get up early to smarten myself up before meeting the new owner and his horse, but he'd have to settle for the dishevelled look instead.

'It's just as well the horses don't care what I look like!' I threw down my flannel, wrinkling my nose. 'At least I'll smell all right!' I said out loud, directing a jet of my favourite scent on to my neck.

Ned was leaning on the warm Aga stove in the kitchen as I shot down the stairs. I'd lived with my grandfather since I was two years old. He was tall and distinguished and very special to me – more like a friend really. Ned was the only family I had, along with our old sheepdog, Shelley.

'Don't you want some coffee? The kettle's just boiled.' Ned's very keen on feeding and watering me!

'Sorry, no time!' I planted a big kiss on his cheek and grabbed a hunk of bread as I ran out to my car.

'I don't think the new owner's going to make much of you, either!' I laughed as I turned the ignition key. 'Come on, Betsy! Please start!'

My newly-acquired car had definitely seen better days. It was old, dented and painted a hideous purple. I didn't care, it was my first car and I loved it. It had been ideal to learn on and very cheap to buy. No-one else had appreciated its awful paint job! But I liked the way everyone stared as I drove through the village on my way to the Mainwaring's yard.

Ben and Sue Mainwaring were in the kitchen as I skidded to a halt on the gravel outside.

'Only three weeks since she passed her test,' snorted Ben, 'and she fancies herself as Nigel Mansell already!'

'Well, I don't suppose it matters,' said Sue, holding up her hands in mock horror. 'With the mess the builders have made, this place already looks as though the house has been used as a racetrack.'

Poor Sue. For years she'd dreamt about having a new kitchen, but the reality was fast turning into a nightmare. Our win at the Cheltenham Gold Cup had provided the cash to do up the kitchen, but the strain of trying to feed the growing number of stable hands in a bomb site, was starting to tell. She was beginning to wish she'd kept her beloved, shabby old kitchen.

'It'll all be worth it, Sue, when you've got everything back in your new pine cupboards. Just you wait and see.' I smiled reassuringly at her.

Sue had tried to tidy things up, ready for the new owner. Edward Holsborough was a successful businessman and his was the sort of money the Mainwarings desperately needed if they were to become top-class trainers. 'If only the workmen had finished this bit, it might make a better impression,' she murmured, sweeping a pile of dirty laundry into the half-built utility room.

Ben gave her ample behind a friendly pat. 'Mr Holsborough's going to be far more interested in the stables than the state of our kitchen!' he laughed.

'I suppose that's why you're wearing aftershave and your best cords today, is it? All the better to muck out the stables?'

'Hey, jump to it, everyone!' I shouted. I could see Mr Holsborough's Mercedes turn into the driveway ahead of the horsebox. 'They're here!'

I felt my stomach tighten as we went out to greet the new owner and the new addition to our yard. Edward Holsborough was one of the country's best known amateur jockeys – one of the privileged few who rode their own horses. He was a tall thin man in his early forties. I'd often seen him on television and although he wasn't an elegant rider, he was certainly effective. He was very tall for a jockey and I always thought he looked rather comical when he was riding. His long arms and legs and his thin body made him look like a monkey up a tree! But in the flesh, he was authoritative and business-like and he expected us to jump at his every word.

Although I was expecting something special, I gasped as the huge grey gelding was led down the ramp. Respect was the first word that came into my mind as I took in his huge size. Paddy's Pet stood 17.2 hands high and was awe-inspiring.

'Wow, he's a monster!' I blurted out, nervously.

Mr Holsborough raised one eyebrow and looked down at me with a disdainful look. 'Hardly. But he's not just one of your ponies, you know!'

I flushed, angrily, but Ben paid no attention to his arrogant remark. He was far too busy running his expert eye over the horse.

'You know, I haven't seen a thoroughbred like him for a long time. In fact he's so full of substance that he almost doesn't *look* like a thoroughbred – if that doesn't sound too Irish!'

Sue laughed, nervously, but I could see she was just as impressed. 'Come and have some coffee, Mr Holsborough, while Becky settles him in.'

I felt very small as I took hold of the halter rope and looked up at Paddy's huge head. 'Hello there, big fella! Welcome to our humble yard!'

Paddy wasn't impressed. He nudged my arm and pretended to bite me, nearly treading on my foot.

'Paddy's Pet indeed!' I snorted. 'You're no pet, my lad. You're really quite bad mannered. But I suppose it would be too much to hope that you'd be fantastic *and* well behaved!'

As we walked through the yard, we passed Red Rag's stable and he put his head out to see what was going on. Red Rag was my real favourite and it was all thanks to him that the yard was doing so well. He was the first horse I'd helped train and he was a star.

I'd never wanted to do anything but work with horses and at sixteen I couldn't leave school soon enough, although Ned had bullied me to stay on and do my exams. I'd gone straight to work at the Mainwaring's yard and I couldn't have wished for a better boss. Ben allowed me all sorts of privileges and when

Red Rag first came to the yard, Ben said I could be Assistant Trainer. The horse was special and I was so proud when Monkey – which was my pet name for him – went on to win a string of races, culminating in the Cheltenham Gold Cup last March.

'Don't worry, Monkey,' I called as we went past, 'you'll always be my number one, even though I've got a feeling this one's going to be rather special, too.'

I led Paddy round the corner of the yard, slightly away from the rest of the horses, as a precaution against infection. I put him in one of the new stables which had been built with the Gold Cup prize money. Ben had extended the stables hoping to attract new customers after our success, and Paddy was one of the first. Mr Holsborough was planning to send us another couple of youngsters for training in the near future.

Back in the kitchen, he was starting to tell Ben and Sue his plans for the coming season.

'I want Paddy to go to the Pardubice in the Czech Republic. He's a big horse and it's just the sort of course that would suit him. That's assuming you do your job properly,' he said, giving me a long, hard stare as I came into the room. 'It shouldn't take you too long to work out that this horse has some problems.'

I sat down and tried to make myself as unobtrusive as possible. He might have loads of money, but I didn't like this man. He needed to tell us about the horse but this sounded more like a lecture.

'He's been bad to start, in fact last season he failed to start, twice. He got in a state when the tape went up. Paddy's got great potential but he's not an easy horse to read, or ride! He'll be quite a challenge for you!'

The Pardubice! I'd heard of the race, of course. It was pronounced Par-doo-bee-chee and was Eastern Europe's answer to the Grand National. But in many ways it was tougher. The huge ditches, together with enormous fences, meant it was far more difficult and imposing than anything jumped over here. It was the ultimate test of horse and rider. To win the Pardubice was the best prize of all and now we had the horse I instinctively knew could do it, even if he did have a few problems. Anyway, I liked a challenge!

'Becky, come quickly, Paddy doesn't look right!'

John was out of breath as he leaned round the door of the hostel that was home to the four stable lads. I lived nearby with Ned, but the others all stayed in the yard, and I often dropped in for coffee and a chat at the end of the day. I was just in the middle of telling Jessie, one of the girls, the latest on Jamie Howland (the love of my life), when John burst in.

I ran with him to the stable and sure enough, Paddy was box-walking, pacing round and round the stable looking very unhappy.

'Do you think he might have colic?' John asked.

'There's only one way to find out,' I replied, looking for a thermometer. I let myself quietly into the stable muttering soothing words to

Paddy as I walked towards him. I put my ear to his stomach.

'That sounds OK.' The relief showed in my face. I took his pulse and waited for the thermometer reading. 'Quite normal. You know, John, I don't think this is a physical thing. He's highly strung and my guess is he doesn't like being on his own. He wants company!'

I led him out of the stall into the sultry summer's night and stood with him in the main yard. I didn't pat him or make a fuss, I just kept talking calmly and after ten minutes or so he seemed more relaxed.

'So, it's company you want is it, Paddy?' I took my first decision as his Assistant Trainer and turned to Jessie who'd followed me down to the stable. 'Can you get the stable next to Flytrap ready, Jess? I think this boy wants to move house.'

It was quite late by the time we had everything organized.

'Are you going to see the Governor now?' asked Jessie. It was the custom in all yards for the lads to refer to the boss as the Governor. I nodded, and she went on, 'Would you tell him all the others have been done, and that the bandage on Dubby's leg can come off tomorrow. Thanks Becky!'

Ben and Sue were watching TV in the small study as I knocked on their door. 'Sorry to bother you, Ben. Can I have a word about the new horse?'

'Come in, Becky,' smiled Sue. 'The Governor's a bit more human now he's back in his jeans! Mr Holsborough left half an hour ago.'

Ben and I discussed Paddy's behaviour in the stable, and I told both of them the thoughts that had been knocking around in my mind. I'd come up with an unusual solution for Paddy's loneliness!

'Well, it's certainly worth a try,' agreed Ben when I'd finished telling him my suggestion, 'and I'd like you to stay here tonight – just to see that Paddy settles in.'

I walked back through the yard to the hostel. Patsy, the new girl, was making coffee. She was a bit stuck-up because her father was one of the owners, and although he'd told Ben to treat her exactly the same as the rest of us, it had set her apart. I went through to the sitting room.

'Tough luck, Jessie! I'm sharing with you tonight, if you don't mind. I'm staying here to keep an eye on Paddy.' We were both secretly delighted. Now I'd be able to finish telling her about Jamie.

'I'll just give Ned a ring and let him know what's happening. I want him to meet me first thing tomorrow.'

I was smiling as I picked up the phone. The solution to Paddy's problem was close at hand!

CHAPTER TWO

The answer was Eric – the goat!

I'd first noticed (or rather smelt) the goat on one of our hacks past the Jackson's farm. Eric was a billy goat and he stank with an overpowering, distinctive smell that really hit the nostrils! I'd heard of difficult horses who'd been put with a sheep or a goat as a stable mate. Apparently, their behaviour had always improved, strange as it seemed.

I'd suggested to Ben that the Jackson's goat could be the friend Paddy needed, and after a couple of quick phone calls the night before, Mr Jackson had agreed to lend him to us, to see how they got on. But getting Eric back to the yard was far from straightforward!

'Becky, for goodness sake stop that darned

creature from eating what's left of my old Landrover!'

I'd managed to persuade Ned to help me and now I was clambering over the seats, trying to pull Eric's head up. 'Oh no! Help! He's having a go at me now, too!'

Eric wasn't too fussy about what he ate, he'd try anything, and now he'd transferred his attentions to the buttons on my anorak.

'How much further, Ned? If we don't get to the yard soon I'll be down to my birthday suit!'

'Well, you needn't worry,' laughed Ned. 'No-one's going to look at you anyway. You stink to high heaven!'

'Thanks a bunch, Ned. Well, I hope *Paddy* likes the way you smell, Eric, or you'll be out of a job. You're certainly not coming to live with me!'

They were an odd couple. But in the days and weeks that followed they became inseparable – Eric sharing Paddy's stable, and eating and sleeping at the same times. Eric would even travel in the horsebox with Paddy to all the major races.

'What news from Jamie, then?' asked Ned, changing the subject.

I could feel myself blushing, but I tried to sound casual. 'Oh, I haven't heard anything since I got my birthday card, but he did say he'd phone sometime.'

Jamie Howland was really gorgeous. He was one of the country's up-and-coming jump jockeys and was two and a half years older than me. He had a lazy smile and dark hair that fell across his forehead – and he and I had got pretty close when he'd been riding Monkey. But now Jamie was in Ireland where he'd been offered a good retainer to ride for one of their most prestigious yards. They'd made the offer after Jamie had ridden Red Rag to victory at the Cheltenham Gold Cup – and it was an offer he couldn't refuse.

In the run up to Cheltenham we'd seen a lot of each other, and it wasn't long before I was pretty crazy about him. I couldn't believe it when he'd said he felt the same! It was wonderful – that was until Jamie had told me he'd accepted the job in Ireland.

I was in a mess. I knew he had to go for his career's sake, but I didn't know whether I could trust him, and deep down I felt jealous and confused. He was bound to meet loads of pretty girls in Ireland. He'd said it wasn't fair to ask me to hang around for him – I was too young, and we should both feel free to see other people.

It wasn't exactly what I'd wanted to hear, but I agreed and threw myself into my work. The trouble was, the talent at the yard was nil. There were just two boys; John, who was solid and reliable, and Andy, who was a bit of a mystery. But neither of them was my idea of a classic hunk!

Ben was walking round the corner of the yard as we finished settling Eric into the stable.

'Ah, Becky. I'd like to see Paddy on his own, without the other horses. We'll go for a quiet hack, just the two of us. Tack up Brian for me, and you can take Paddy. Let's see what he's really like.'

I had to have my wits about me as I got Paddy ready. He really was annoying! One minute he was disdainful and aloof, not really interested in human contact, and the next he was biting and nudging, to try and attract my attention. I tried not to hurry but I couldn't wait to see what he was like – and I wasn't disappointed.

First impressions were thrilling! I'd ridden some big horses in my time but this one was enormous! I felt as if I was towering above Ben and, as we set off, I was aware of the powerhouse beneath me. I had to remind myself to concentrate hard in case he came up with any of his tricks.

Ben looked over his shoulder. 'What about a short trot? How does he feel, Becky?'

'He's amazing!' I shouted, as I urged Paddy forward into a trot. 'He's got the most incredible movement! It's like being on a springboard!'

'Go on, then!' called Ben, reading my thoughts. 'You can have a quick canter, just to see what he's like!'

He reined in as Paddy and I went up a gear, making a smooth transition into a canter. His enormous stride ate up the ground and there was an overwhelming sense of power. It felt like riding a cruise missile! A huge smile spread over my face. This horse was really special. He'd either be brilliant or useless – but never mediocre! I wondered what his gallop would be like but, tempted though I was, I knew I'd have to wait until he'd progressed through weeks of structured training. It would be something wonderful to look forward to.

'Woah there, Titch!' I said, bringing him back to a trot. Ben looked on with admiration from the far corner of the field and, as we approached, I leaned forward over Paddy's neck and gave him an appreciative pat.

'Who's a good boy, then!'

But I was totally unprepared for Paddy's reaction. He laid back his ears and kicked out, violently.

'Steady! There's no need for that!' I kept my voice calm and managed to regain control.

'What was all that about, you silly thing! I was just giving you a pat.'

'Well, that's *one* problem, Becky.' Ben frowned. 'I wonder how many more bad habits he has. We'll need to find out why he reacted so strongly.'

As Ben and I hacked slowly back to the yard, my brain was already in overdrive, assessing Paddy and his problems and working out his training plan. I had to get him fit for the big race in four month's time!

'Supper's ready! Come and get it.'

Ned placed the most wonderful-smelling stew in the middle of the kitchen table. 'There you are, Becky. Your favourite!'

The kitchen at Nettleton Lodge was stylish but cosy and the flag floor was polished to a fine sheen by Mrs Weaver, who came in twice a week. Ned had a dining room for his proper entertaining, but when we were on our own, we liked to eat in the warm kitchen at the big oak table with its print tablecloth. Ned was an amazing cook and I felt very lucky to have him as a grandfather.

I couldn't remember my parents – I was only two years old when they were killed in a car crash, and now Ned was the only family I had. He'd given up his career in the SAS to look after me. 'With your sweet fair curls, you looked just like your mother did at your age,' he used to say. 'How could I resist you!'

Ned had led a pretty glamorous life, really. He'd travelled all over the world in his military career. Most of his work had been top secret, but occasionally he'd let details slip about his contacts with foreign royal households and dangerous skirmishes in the desert. He still kept in touch with some of his old chums in Whitehall, and I was under the impression that the government consulted him occasionally on military strategy. Not that he would ever say anything about it.

'There's some chocolate fudge cake – or yoghurt and fruit if you're on a diet again!'

said Ned, as I pushed my plate away and sighed contentedly.

'Healthy eating, Ned, actually!' I pretended to be indignant as I leaned across to the fruit bowl. 'We haven't all got fabulous figures like you! Is there anything good on the box tonight?' I changed the subject quickly.

'Mmm. What's that?' said Ned, distractedly. He was reading an airmail letter. 'Claudia sends her love. Oh! And she says to thank you very much for the latest report on Florida's Hopes.'

Claudia came from Florida. She'd first met Ned years ago in the States when he'd gone to play in a polo match organised by her oil-baron husband. They'd exchanged Christmas cards ever since. Claudia had written last year to say she'd been widowed and was coming over to the UK for a visit and would it be possible to meet up.

Claudia was very beautiful – blonde, with an all-the-year-round tan; larger than life and very rich. But much as I'd enjoyed her visit, it had come as quite a shock when they'd started talking about marriage!

Florida's Hopes was a horse that Claudia had bought on her last trip. She'd put it in training with the Mainwarings and I used my computer to keep her in touch with its progress. It was something I did for all the owners. Sue thought it looked really impressive.

'When's she thinking of coming back?' I asked casually, as I curled up on the rug and gave Shelley a hug.

'She's been held up with a couple of things and needs time to tie up a few loose ends.' Ned scanned the letter. 'Listen, this is typical Claudia! She says it was lovely to get back to the American shopping malls and supermarkets – and to tell you that her hairdresser, Fabrizio, was horrified that the salons here hadn't kept her streaks in the manner to which they were accustomed!'

I liked Claudia a lot but I was secretly relieved that I would have Ned to myself for a little longer.

'Tell her I hope she manages to tie up her loose ends – and split ends!' I added, cheekily.

CHAPTER THREE

There was a sharper edge to the air as the days shortened and the countryside seemed to pause, as if getting ready for the autumn weather. The grass wasn't growing as fast, and I shivered momentarily as I grabbed my body warmer from the back of the car.

Paddy and I were now locked into a comfortable routine, and he was feeling good. His body had muscled-up after the weeks of training that I'd specially tailored to his needs, taking him back through some of the basics to deal with his problems. I whistled as I rounded the corner of the stable block and I knew Paddy was listening for me.

'On your best behaviour, Titch. And no biting today!' I threatened playfully, as I pushed him to one side and prepared to lead him out of the stable. 'See you later, Eric. You know, I don't think you smell too bad today – or maybe I'm just getting used to you!'

Paddy liked to be in the lead, and we took our place at the head of the first lot, the name given to the first group of horses. We were getting ready to go out on exercise before starting their programme of work for the day.

I swung round in the saddle to check that everyone was ready. 'OK, team, let's go past Kiddington today.'

We tried to vary the hacks, and had about five routes we regularly used. Horses, like humans, appreciate a change of routine. We hadn't gone far before a sporty, dark blue BMW pulled up ahead of me, and the woman driver asked how to get to the village of Lower Nettleton – there were loads of expensive antique shops there. She didn't turn down her music, and I was proud of the way Paddy stood quietly while I told her the way. I pointed my whip in the direction of the church spire over the hill when suddenly, without any warning, Paddy reared up in great distress, and started backwards into the rest of the string.

In a blind panic, his hooves beat out a jerky rhythm in the air, narrowly missing the woman who was leaning out of her car window. I don't know whose face looked worse – hers or mine! We were both terrified! It was like being on a roller-coaster as I struggled to regain control, aware of the other horses behind, and the nearness of the perfect, shiny new car. My stomach lurched and adrenalin surged through my veins. But I sat tight, as Paddy snorted and pranced sideways, and eventually came back to a nervous stand.

'Aren't you rather young to be in charge of so many highly-strung horses?' shouted the driver, as she accelerated away from the string, now in total disarray.

'Are you OK, Becky?' Jessie was the first to speak. All the lads were wary of the big horse and this confirmed their feelings. Secretly, they thought Paddy was a bit loopy and too powerful for a girl like me. Not that they were offering to do the job! It took me a minute before I could answer.

'Sorry, gang. That was nasty, but we're OK now. Let's carry on.' I took charge of the string again.

For the rest of the ride I puzzled over Paddy's behaviour. What was wrong with him? First, there'd been the problem when I'd patted him, but I'd had some success working on that one with every rider's secret weapon – polo mints! At the end of each day's work, I would hold a mint in my hand, as I slid my arm along Paddy's neck to pat him. Gradually, he'd got the message. It was great what a bit of bribery could achieve!

But this problem with the whip was a new one. Then it came to me; I remembered what Mr Holsborough had said on the first day they arrived. Paddy had failed to start at his last two races. Could it have anything to do with the whip? I decided to experiment when we returned from the morning's schooling.

'Andy,' I said when I'd dismounted back at the yard. 'Can you hold Paddy's head for me?

I've got an idea.'

I simulated what I thought might happen at the start of a race. The jockeys would all be bunched up, and often they would have to kick the horses on and put their arms up their necks, to encourage them forward to the start line. And – yes! As my arms went along Paddy's neck, the whip angle changed. That would seem threatening to any horse that had been mistreated. And Paddy *had* been mistreated, I was sure of that now! Confirmation was instant. Paddy did an action replay of the morning's rodeo antics – but this time I was ready for him!

'Eureka! Thanks, Andy. At least now we know what's causing the problem.'

Sorting it out would be another matter, but I'd already decided that, from now on, I would ride Paddy without a whip. I'd feel a little bare, but it'd be worth a try.

Ben was waiting for me in the yard as I led the

second string back in. 'There's a meeting in the kitchen in half an hour, Becky, if you could make it,' he shouted. 'Oh, and take these! You might need them tomorrow. I hear you're not old enough to be in charge!'

I blushed furiously as I caught hold of the L-plates which he tossed in my direction. That woman with the BMW must have phoned Ben to complain. The old bag! I was tongue-tied for a moment as I tried to think of something clever to say.

Ben winked reassuringly. 'She soon shut up when I told her you'd trained the Cheltenham Gold Cup winner, no less!

'Thanks, Governor!' I beamed with relief as the others burst out laughing. 'I don't really blame her, though,' I added generously. 'For a moment it was more like the Grand National out there. Paddy nearly jumped over her car! See you in a minute!'

Back in the kitchen, Sue sighed as she leafed through her battered diary and the pile of papers strewn all over the table.

'It's no good, Ben, we're going to have to

build an office. Nowadays there's far too much important paperwork passing through the yard to risk losing anything. I was looking for a recipe the other day and I found the Aintree entry forms inside the cookery book. That could have been serious!'

'I thought Becky was helping with some of the admin. on her computer.' Ben scratched his head.

'Hang on a minute, you two,' I butted in, as I pulled out a chair and sat down. 'Sorting this lot out will take a miracle, not just a computer! But I agree with Sue, something definitely needs to be done.'

'OK, point taken!' said Ben, suddenly looking tired. 'The yard expansion has rather crept up on us, and we've not been organized. Owners like Mr Holsborough are used to a more professional service, and I suppose we'd better do something about it, otherwise we'll never keep up with the horses let alone the owners!'

It was agreed that Patsy would spend a couple of hours a day entering the details of the horses' racing fixtures and feeding requirements into the computer. The small study would be converted into a proper office and I suggested that Ben and Sue buy a fax machine, while they were at it. I was sure Patsy would be delighted with the alteration to her day. At least indoors her nail varnish wouldn't get chipped! It would give the rest of us a break as well from hearing how Daddy was going to set her up with her own yard. As soon as she'd learnt everything from us, that is!

Our meeting was to discuss arrangements for our trip to the Czech Republic in just four weeks' time. I could hardly believe it! Travelling to any race meeting was a lot of work – preparing for a trip abroad would be exhausting. Riders had little to worry about, all they had to do was turn up and ride. But it was my job to make sure that everything else was organised. I had to iron the silks, pack the racing saddle, tack and boots, and even take a spare set of shoes for the horse. In fact, packing for the horses was the major job. They even had their own passports, and we always took our own food and water for the horses so that they didn't have to cope with a change. Horses can go down with jippy tummies too and, on top of all this, I had a goat to pack for as well!

The horsebox was parked away from the stables by the side of the old pig shed. Some yards had enormous boxes, bigger than most people's houses – *and* more luxurious! Ours was a bit on the small side, but there was room for three or four horses, and our quarters had a shower, loo and small kitchen, complete with microwave and fridge!

There was a musty smell as I climbed into the living area. 'Time for a spring clean before the new season starts,' I muttered, as I pulled an old packet of soup from one of the cupboards. 'I wonder what state the fridge is in – probably green and mouldy.'

Jessie hurled herself round the door when she heard my shout. 'There's a huge creepy crawly in there!' I gasped, pointing a shaking hand at the fridge. 'It's disgusting!'

'You great wally!' laughed Jessie, as she extracted the offending creature between her finger and thumb. 'It's one of Patsy's false eyelashes! What on earth are they doing in there?'

'Oh, that's nothing!' I replied, relieved now the danger was gone. 'When she's travelling with the horses she even keeps her lipsticks in there – Christian Dior, of *course*!'

'No wonder she's so sickeningly thin,' laughed Jessie, 'if that's all she puts in the fridge. What a great diet! It beats those Lion Bars you throw in.'

I closed the door behind us. I'd start on the horsebox tomorrow, after I'd drawn up a check list. It was a big responsibility, but if we got it right, it could make the difference between success and failure.

Sue was coming out of the feed room and flagged me down as I crossed the yard.

'If you have an hour this afternoon I'll run through Paddy's feed chart for his trip. We'll need to start reducing his protein soon.'

We always cut back the protein when a horse had a long journey ahead. Paddy wouldn't be getting much exercise en route so he couldn't burn off the calories. In fact, there was a danger he'd get 'tie-up' – a kind of muscle cramp, if we didn't reduce the protein.

Sue was the feeding expert. She believed in routine and insisted that it was just as important to stick to a feeding regime as it was to follow a working regime. That was another reason we took our own feed with us – so we were in control of what they ate.

Patsy sauntered over as I made my way across the yard. As usual, she looked immaculate. Whenever I swept the yard, I ended up looking a mess, but Patsy had hardly a hair out of place and her jodpurs were still quite clean. She had a copy of the *Mail* in her hand.

'Hey, Becky, I think there's a friend of yours on the gossip page today. You know Jamie Howland, don't you?'

She thrust the page towards me. There he was, looking as gorgeous as ever. But who was that hanging on his arm? My stomach lurched. There, gazing up at him with adoring eyes, was that cow, Sophie Redmond! I knew her when she worked at the Mead's yard down the road, but when the yard closed, she'd vanished from the scene. I thought she was working at Lord Clarendon's stud in Gloucestershire. So what was she doing in Ireland – and what was she doing with Jamie?

'Oh, yes,' I said, quickly. 'Jamie told me he needed a partner for the Charity Ball and that Sophie was going with him.' There was no way I'd give Patsy the satisfaction of seeing how hurt I was. 'Sophie's an old friend of mine, too,' I added. 'See you all tomorrow, then. I've got stacks of paperwork waiting for me at home.'

I licked away the salty taste as a tear rolled slowly down my cheek. I slammed the car door and slowly put her into gear. 'Let's go Betsy! He's not worth it.'

CHAPTER FOUR

I knew Claudia was back the minute I opened the Lodge door – her distinctive scent hung in the air. I tried to look cheerful as I kicked off my boots and went into the kitchen. Claudia was standing at the Aga with her arms wrapped around Ned.

'OOPS! Caught in the act!' she grinned. 'Becky, how are you? You look great!'

She wrapped me in her long tanned arms and gave me a big hug and, in spite of myself, I had to smile. Americans were something else! They had none of the British stuffiness and Claudia personified their easy going charm – and was attractively eccentric as well!

'It's just great to see you, Becky. I can't wait to catch up on your news. And I want to hear all about the new horse!'

She held me at arms length and looked me up and down. Next to her, I felt scruffy and dirty.

'You probably want to freshen up before supper so, on your way to the bathroom, you might want to look at the gift I've brought you. It's in your bedroom.'

Claudia's presents were as extravagant as she was, and I shot upstairs, desperate to see what she'd brought. As I ran into the room, something hurled itself at me. I skidded on the rug, crashing to the floor, and struggled to wipe the wetness off my face. Then my eyes focused on a long-haired, tan and white Irish Terrier, busy trying to lick me all over!

'A dog!' I yelled, as I scooped him up and ran downstairs. 'Oh, Claudia, is he all mine?'

She and Ned were laughing. 'All yours,' they confirmed, 'but you'll have to give him a name.'

'Oh, dear!' I lifted him up and looked him straight in the eyes. He cocked his head to one side.

'Scruffy little devil, isn't he?' said Ned.

'Well, I didn't think a pedicured Pekinese was quite Becky's style!' Claudia winked at me. 'I thought she'd prefer the scruffy look.'

'That's it!' I laughed, hugging the dog. 'You're certainly scruffy, so Scruffy it shall be! Can he sleep in a basket at the end of my bed? Oh, please say yes, Ned!'

'Well, yes, I suppose so,' Ned laughed happily. 'But he's not allowed *on* the bed!'

I threw my arms round them both, and the puppy yelped as Shelley looked on with a resigned air. She could sense her life would never be quite the same again!

Now that Paddy was really fit, Edward Holsborough was spending more time at the yard. He usually managed two or three sessions a week, driving down from London as much as his business interests would allow. Today, Paddy was going up the schooling fences on the gallops, and Ben and I sat in the

Landrover and watched.

'I still can't get over how awkward he looks.' Ben adjusted his binoculars, focusing on Edward's long thin legs, bent double in the short stirrups.

'Strangely enough, he's not cramping Paddy's style,' Ben observed. 'That horse is certainly jumping better! You've done a great job, Becky!'

Before he came to us, Paddy had finished fifth in the Foxhunters at Aintree, which wasn't bad by anyone's standards, but he was performing better now. 'Let's hope Edward appreciates what you've done,' Ben added.

'Steady on, Governor,' I laughed, 'that almost sounds like praise!'

'Well, you needn't think of asking for a pay rise, young lady! We haven't had any results – yet!' Ben grinned at me.

Mr Holsborough was sweating as he pulled Paddy back to a walk. His speech was short and jerky as he struggled for breath.

'My God, he's still a bit of a handful! Sometimes it's hard to tell who's in control!

He's so strong, it's as though he's *driven.*' As usual, Mr Holsborough barely looked at me, directing most of his remarks to Ben. It was really infuriating. He knew I was doing most of the training but it was as if I didn't exist.

'Paddy must like her,' he nodded in my direction, 'otherwise she'd never be able to hold him.'

I grinned at his back-handed compliment. Although he was arrogant, Edward Holsborough had exactly described the sensation of what it felt like to jump Paddy. I only wished *I* could compete at the Pardubice!

'OK, back for breakfast.' Ben's voice interrupted my thoughts. 'I bet Mr Holsborough's worked up quite an appetite!'

'I've got an appetite, all right! But I hope your lovely wife doesn't put a cooked breakfast in front of me – I'm trying to *lose* weight, you know!' He patted his stomach. I groaned, knowing full well that Sue was preparing an enormous fry-up. When he saw it, no doubt Mr Holsborough would be his usual charming self!

Scruffy and I became inseparable in the week or so before we left. He was such a character. He slept in my bedroom – not always in the basket, I have to admit – preferring to burrow into the duvet to get the warmth from my toes. He was just what I needed, something to shower with masses of affection. I was really missing Jamie, even though I didn't want to admit it.

Ben said it was OK for Scruffy to come to work with me, and it only took one well-aimed kick from Paddy, before Scruffy learnt to keep clear of the horses' hooves! His favourite place was in the tack room, especially when the heater was on, and whenever I was working around the stables he'd be curled up on a bale of hay, often snoring in time to the pop music blaring from the radio! The only area where the dogs weren't allowed was the arena, especially when we were using it for a horse like Paddy who needed special attention.

Scruffy followed me as I made my final inspection of the horsebox. I ticked off the last item on the clipboard, and checked that Edward's silks were hanging in the wardrobe. The fridge was full, and tucked away under one of the seats were a couple of bottles of champagne, just in case we got lucky.

Jessie peered round the door, admiringly. 'I can't believe how well organised you are, Becky. It looks immaculate!'

'Don't you believe it,' I joked. 'I've probably forgotten the loo paper or something just as important! Oh, no! I can't believe it!' I shrieked, as I peered into the small loo. 'I *have* forgotten it! Ned would have killed me! And I bet they don't have the soft stuff over there!' I made a note on my clipboard.

It was then we heard the most awful scream.

'Quick! It's coming from the hostel!' cried Jessie, peering round the lorry door.

We clambered out in time to see Patsy in her bathrobe, with a towel wrapped round her head, tearing out of the cottage. Her face was streaked with bright red and orange.

'Who did this?' she wailed, tugging a strand of her once-blonde hair. 'I swear I'll kill whoever did this to me!' She was shouting and waving her bottle of blonde highlighter. 'Someone's put dye in this! Disgusting orange henna!' Patsy stamped her feet and threw the bottle at John, who happened to be nearest and was doubled up at the sight. Jessie's mouth hung open as Patsy swung round, her hair dripping the frightful colour, screaming and swearing at the top of her voice.

'Poor you! How awful,' I spluttered, putting my hand over my mouth. 'And your hair was so beautiful,' I observed, innocently.

Jessie turned to me in horror. 'Becky, you didn't?' She looked shocked.

'Who me?' I said, appearing stunned. 'Why on earth would I do that to Patsy, when she's such a good sport?'

Jessie raised one eyebrow and shook her head in disbelief. 'Well, I'm glad it wasn't me! Patsy looks ready to kill someone! That reminds me, we've decided not to get dressed up tonight. Wear something casual, instead. See you at eight.'

We were having a farewell barbecue before we left the next day, and we'd invited a few friends and the lads from the neighbouring yard.

'OK. I'll come a bit early to help with the tables. I hope Patsy gets her hair sorted out!' I shouted after Jessie, smiling as I ticked the final item off my clipboard.

Ned was pouring over the map when I got back to the Lodge.

'Ah, there you are Becky! I was just making some final checks and I think the route looks fine.'

Ned and I would be travelling together, with Paddy, Eric and the horsebox, and Edward and Ben would fly over later in the week. The journey would take us four days, going through three countries and covering the best part of one and a half thousand miles. I'd never been that far before and I was slightly apprehensive. But Ned was in his element!

'We'll go through Frankfurt and cross into The Czech Republic right here.' He stabbed the map with his pencil. 'Then it's a quick hop to Prague and eastwards to Pardubice – easy!'

Claudia laughed as she came into the study. 'God help you if you leave your reading glasses behind like you normally do.' She winked at me. 'You know, he's as blind as a bat without them!'

'I'm more worried the border guards might find it hard to believe an old-age pensioner's got an HGV licence!' I giggled.

Ned was still fit and attractive and didn't look his age so I couldn't resist teasing him.

'Out of here, you two!' he pointed at the door. 'I've got important planning work to do! And Becky, if you're going to that party tonight,' he retaliated, 'you'd better make sure you're up and fit for duty tomorrow!'

'Yes, Sir! Major!' I saluted, as Claudia and I scurried back to the kitchen.

'And don't you worry about Scruffy,' Claudia smiled, as I hugged my dog. 'I'll take very good care of him, I promise. How about this? I'll even let him come in the Audi with me – but he'll have to keep his paws to himself!'

Scruffy had a weird travelling habit. He liked to balance on the back of my seat with his paws on my shoulders as I drove through the country lanes. I hadn't a clue how he managed to do it, but no matter how hard I tried, he wouldn't sit on the back seat like any normal dog.

Claudia's offer to take him in the car was very generous but it was bound to end in tears. Somehow, I didn't think it would be too long before Scruffy's muddy paws ended up on Claudia's expensive shoulders!

'I'll miss him,' I said, and I was surprised to hear myself murmur, 'and I'll miss you too, Claudia.'

CHAPTER FIVE

The early morning breeze stirred the brightly coloured bunting that decorated the full length of the stable courtyard. I gasped as Ned and I drove over the gravel and came to a halt at the horsebox. Along its side hung a banner with the words 'Good luck, Becks!' The lads all cheered as they saw my stunned expression.

'You toe-rags!' I exclaimed. 'You can't have had any sleep last night!'

'Well, we needed to do something to warm up after our unexpected cold shower!' John laughed.

When I'd arrived to help set up the barbecue the night before, the others had already been hard at work and the barn was looking great. A big space heater was roaring in the corner and Andy was watching the barbecue which Ben had set up near the door. Sue was ferrying over armfuls of cakes and puddings which she carefully set out on the trestle tables.

'You're not going to starve,' she'd announced, wrinkling her nose appreciatively. 'Mmm, those sausages smell great!'

Just then, I'd caught sight of Patsy, laden with even more cakes, heading for the table, and I couldn't help giggling. She looked dreadful! Her hair was a violent orange mess!

In one swift movement, she grabbed a cake and planted it smack in my face.

'Take that!' she said, triumphantly. 'That's for messing with my shampoo!' She'd ground the chocolate sponge well in, just to emphasize her point.

'Oh really?' I reacted instinctively, grabbing the nearest jug of punch. 'Well, by the look of it, there's still some shampoo left in your hair! Allow me to rinse it off!' So saying, I tipped the whole jugful over her carroty thatch. 'And what's more,' I squinted through the chocolate, '*you* could do with a face pack, too!'

Unfortunately, Patsy ducked, and John got the cake instead, smack in the chops! By this time, everyone was shrieking and they all pitched in to the battle! And I have to tell you, it was wonderful! Secretly, I'd always wanted to lob a custard pie in someone's face, and it was everything I'd ever dreamed it would be! Even Sue joined in!

The floor was a slithering mess before we'd stopped, hysterical with laughter. But there were still plenty of cakes left and the damage wasn't total!

'Quick!' I'd shouted, grabbing a hosepipe. 'The others'll be here soon! Let's get rid of the mess!' Thank goodness we'd been in one of the outbuildings – the concrete floor wasn't a problem to clean.

John had grabbed a second hose and had looked around for somewhere to start.

'You've still got some icing on your forehead, Becky!' he noticed, grinning. 'Better get it off before it sets too hard!' With military precision, John had turned his hosepipe on me and I'd screamed as he drenched me with the freezing water!

'You filthy rat!' I'd shrieked, the fight starting, once more. 'But you've picked the wrong person this time! I was brought up on hosepipe fights and I warn you – I'm good!'

I'd grabbed John's waistband and stuck the pipe down the front of his trousers. 'Take that, you miserable coward!' I'd shrieked with laughter as the water poured down his legs.

'And you can take that smile off your face too, Jessie!' I'd grabbed a bucket which I used as a shield as I drove her backwards towards the door.

'Don't think you can hide from me!' I'd yelled, turning on the others who were cowering behind an upturned trestle table. Just then the water spluttered and died to a trickle.

'That's enough high jinks for one night!' Ben was turning the water off at the tap. He'd pretended to be stern but I could see the twinkle in his eyes. 'Sue, I'm surprised at you!' The corners of Ben's mouth began to twitch as he stared at his dripping wife. 'Come on! Hurry up and get changed – the last batch of sausages is going on now!'

In a few minutes, we'd got the barn back to normal and then we'd queued for a hot shower and Jessie'd lent me some clothes for the party. It'd been great! The music had been on full blast and we'd yelled at each other, letting off steam after the weeks of hard work. I'd finally got to bed just before midnight – and then the others must have put up the flags.

They grinned as I admired the yard. 'Thanks a lot, team! It looks great!' I felt a bit weepy – they'd really gone to town! 'I just hope we'll live up to all this.'

Ben and Sue came round the corner. Sue was waving a bright red folder.

'Paddy's passport and his permission to travel.' She handed me the Ministry of Agriculture paperwork. 'And Eric's! Can you believe it? They've even got a department that deals with goats travelling abroad. For goodness sake, don't go without them!'

'No chance!' I laughed. 'I was just coming over for them after I'd said goodbye to Monkey.'

I ran round the corner of the yard. Monkey had his head out of the stall, dying to see what all the fuss was about. His ears twitched as he saw me. I swore he used them like radar, and could almost talk to me!

'Be a good boy, Monkey.' I stroked his soft muzzle gently. 'It won't be long now till you're racing again, and then *you'll* have pride of place in the box!'

I felt strange as I walked away. I'd never left him before.

Ned was whistling as we left the port of Zeebrugge. Thankfully the crossing had been calm – horses can't be physically sick so they find a rough sea very distressing. The lorry was wide and it took me a minute to adjust to driving on the other side of the road. I felt exposed and I dived to the right of the passenger seat, nearly ending up in Ned's lap, as a big Citroën sped past with only inches to spare.

'Crikey, Ned! That was close! Remember we're on the right-hand side here!' I grabbed a tissue to mop up the drink which had spilt out of my can.

'Relax, young lady. May I remind you that I'm the man who negotiated some of the worst terrain in Oman, *and* successfully led his command through, I might add!'

I smiled. It took Ned back to his army days – in fact, he'd planned the trip like a military operation.

'It's modern day Belgium and Germany we're going through, Gramps, not war-torn Arabia forty years ago!'

'Laugh if you like, Miss Clever Clogs, but my knowledge of Czechoslovakia, as it then was, will stand us in good stead, just you wait and see!'

We were well into the open countryside when suddenly he swerved.

'Darned fox! Did you see it, Becky? Hang on, we've hit a tractor rut on the verge!'

Ned wrestled with the steering and skilfully applied the brakes. The sheer weight of the horsebox gave it a momentum of its own, and it seemed like an age before the lorry finally juddered to a stop. It leaned uncomfortably to the right, and we saw the tractor rut had developed into a serious ditch at the side of the road.

I heard Ned curse under his breath and I almost felt like crying, partly from shock and partly from relief that we were OK.

Ned put an arm round my shoulder and gave me a squeeze. 'Why don't you take a look in the back to make sure the boys are all right and I'll see about getting us out of this ditch.'

Paddy and Eric seemed unconcerned by the whole affair, and after I'd checked them over, I joined Ned in the ditch below the truck. He'd already found the wooden blocks we always packed for just such an event, and he strained as he put the final one in place.

'Just a jiffy, and then we'll have her level again,' he grunted.

Ned was under the side of the lorry now, struggling to get the last block in place. And above him, the animals' pee was draining to one side of the box and building up like a dam fit to burst! And it did! As Ned struck the final hammer blow, the pee started to pour down his neck from one of the vents above his head.

It was the last straw! Ned turned a dark purple and I looked on, horrified. But as much as I tried not to, I soon collapsed, helpless and giggling. The tension was released as both of us fell, hysterical with laughter, into the ditch! Finally, Ned grabbed an old towel from the cab and cleaned himself up as best he could.

'That'll have to do until we reach our camp site tonight,' he said grimly, and we settled down once more into the rhythm of the road. We'd only gone a few miles before my nose started to twitch. The heater was on full blast and Ned was drying out fast, but as he did, he started to smell more and more awful!

'Yuk! You smell worse than Eric, Ned!' I wound the window down as fast as I could. 'Make haste to the camp site! And don't spare the horses!'

We were seventy miles from the German border before Ned got a shower! My poor grandfather – he'd never live it down!

The days and the miles merged into each other as we settled into a comfortable regime. We were always on the look out for lay-bys or quiet country lanes where we could pull up and give Paddy some exercise. Paddy was extremely fit and if he didn't work-out en route, he'd rapidly lose condition. It was a long journey but he soon got used to life on the road.

Ned shouted a cheerful goodbye to the border guards as we cleared the Czech passport control.

'Well, that was relatively painless!' he observed. 'In the old days, we'd have had the Czech secret police breathing down our necks, thinking we might be spies!'

'You probably were!' I retorted.

'That would be telling!' he grinned, tapping the side of his nose with a finger.

'Anyhow,' he continued, 'we've established that everyone's friendly now, so let's see what's happened to the countryside since the communists left.'

We'd only gone a few miles over the border and we'd noticed the difference. Germany had obviously been rich and prosperous, but as we drove through the Czech towns, we could see that the country was starting to rebuild itself after years of neglect. I felt a great sense of history all around me.

'Do you think we'll have time to explore Prague after the race? I've heard it's really beautiful.'

Ned sighed. 'Let's get the race over first!' He accelerated past an old Skoda car and grinned. 'That one's definitely past its sell-by date! With a bit of luck we should make the Pardubice racecourse by teatime.'

Ned put his foot down for the last stretch of the journey. We were on the Prague bypass and I could see the river and spires of the old town tantalizingly close as we circled to the East. We turned on to the E67 and I saw my first sign to Pardubice.

'Look Ned, just 140 kilometres to go! I

wonder who else will have arrived? Do you think anyone will be there to meet us and will the stables be OK?'

'Hang on, young lady, you'll find out soon enough. Now change that tape and put something decent on!'

I shoved in the Bach Violin Concerto – that'd keep him quiet – and walked through to the back of the lorry. Paddy and Eric seemed blissfully unaware of all the excitement.

'Guess what, Paddy?' I shouted. 'We're almost there!'

CHAPTER SIX

'Oh, my legs! They're all cramped!' I groaned, as I slithered down from my seat and jumped out of the cab on to the grass below. We had arrived at the racecourse on the outskirts of the town. A magnificent forest ran round one side of the course and I squinted at some of the fences in the distance. There was tight security in the stable area and a guard pointed to where we should park.

Suddenly, I felt very tired and dirty. 'I could die for a good soak in a bubble bath. The last part of that journey was a killer.'

I knelt down to adjust my laces and as I stood up, I came face-to-face with the most gorgeous blond hunk, clearly having difficulty with his English.

'Good night!' He said politely and pointed to the side of our lorry. 'I see you are coming from England. Welcome to Pardubice!'

'Oh, thanks.' I said quickly, putting out my hand. 'I'm Becky Johnston and this is my grandfather, Ned Hunter-Gordon. It's great to be here, I can tell you. It's been a long drive.' I tried not to seem too flustered – his eyes were a brilliant blue! 'And you are... ?' I inquired.

'Tomas. Tomas Stravinsky. At your service!' Tomas told us he was also riding in the Grand Pardubice on a talented eight year old called Genesis. It turned out Tomas had actually heard of Edward Holsborough and was quite impressed when he saw Paddy.

'He looks very good, that horse, extra good in the backside!'

Ned and I smiled at each other. 'Strong in the hindquarters,' I corrected, and grinned at him.

Tomas followed me as I turned back into the horsebox. Suddenly, he let out a gasp of surprise as he tripped over, falling into the straw.

'What is *that*?' he shouted; it was Eric, nudging him and chewing at his jacket.

'That's only Eric!' I raised my arms in exasperation. 'Your manners leave a lot be desired, you wicked old goat. May I introduce Tomas? You'd better be nice to him – he's a friend!'

'A goat!' Tomas blinked. 'And you bring him all the way to Pardubice?'

'We wouldn't leave home without him,' I laughed. 'Paddy would be most unhappy if we left Eric behind!'

Tomas was a huge help as we settled into the stables and found our way about. He acted as our interpreter – and a very dishy one, too. We quickly became friends and by the time the other competitors started to arrive, the Mainwaring camp was well-established.

We weren't allowed on the racecourse yet, but an adjacent field, with good going, had been allocated for the horses. We'd allowed a week to get Paddy back into tip-top condition and gradually I increased his workload, holding him back to a fast canter. We'd only let him gallop on the day before the race, on a short workout over some of the fences on the course. This was extremely unusual – I couldn't think of any other country where you got a trial run over the fences before the actual race.

Everyone admired Paddy, and I must say he looked brilliant! It seemed as if my training plan had been a success but we'd soon see whether his performance lived up to his good looks.

Ben and Mr Holsborough flew in three days before the race, and Ned and I went to meet them. On the two-hour drive back to the stables, I filled them in on all the news, telling them how brilliant Tomas had been. We'd been invited to have tea with Tomas and his father, and as we had to wait another day before we could look at the fences, we all thought we should go. From what little Tomas had told me, his father sounded very interesting!

'Wow, Tomas!' I gasped, as we swept up the

elegant tree-lined drive which led to Tomas's house. 'This isn't a house, it's an ancestral mansion!' I glimpsed the house through the trees as we followed the bend in the drive past an old boating house by the side of the lake.

'It is a mansion, you are correct!' he laughed, 'a very expensive mansion! But I did not grow up here, you know.'

Tomas explained how the old house had been their family home until it was seized by the State after the Second World War. They'd only recently got it back, and Tomas's father, a well-known biochemist, ran a research laboratory in the grounds. Professor Stravinsky was investing the profits from the lab back into the property and now they were beginning the expensive job of restoring the house and its grounds to their former glory.

'It'll take a lot of Koruna – lots of dosh, I think you say!'

Ned and Tomas's father hit it off immediately; after just five minutes they discovered they had several acquaintances in common. As the two men chatted, Tomas and

I took off for a quick tour of the grounds. The stable block was amazing. It was set in a beautiful cobbled courtyard and an archway lead through to the old wooden stalls.

'Tomas, it's beautiful!' I breathed.

'It will be the last part of the property to be restored,' said Tomas sadly, but his eyes were bright and full of resolve as he added, 'but one day, these stables will once again house the magnificent Kladruby horses for which the Czech Republic is very famous.'

I'd seen pictures of the Kladruby – they were huge carriage horses, either ebony black or grey.

Tomas continued: 'We have only one of them now but soon there will be more!'

I was especially interested to see the horse after Tomas had told me his father was using blood samples from it, to develop a new equine serum. We caught up with the others in the newly-built laboratory, a short distance behind the house.

Tomas's father beamed as we walked in. 'Miss Becky, may I present my assistant, Pavel

Husak.' Husak looked about forty – a rather short man with lank, receding hair and deep set eyes which never seemed to leave me as I walked across the lab to meet him.

'Delighted,' Husak said, bowing slightly as he shook my hand, and held it for rather too long. 'Tomas has told us all about your powerful Irish horse – and a bit about you, too.' The others laughed at Husak's attempt at a joke, but I thought there was something sly about him and I pulled back my hand, pretending I hadn't noticed. His English was excellent with traces of an American accent from his post-graduate years in the States.

The Professor continued: 'I am just starting to show the others the work we do here, if you'd care to join us?'

It was fascinating. The Professor and Husak were involved in some major scientific work, developing a serum against a lethal virus – a rare strain of African Horse Sickness. We'd heard about the illness when it hit the headlines as it swept through Australia. There'd been dreadful stories on the news about how the virus attacked the horses' lungs and we'd seen pictures of horses dying. Often they were left weak and diseased, their racing days over for ever.

Mr Holsborough, a typical businessman, wanted to know more, and was grilling the Professor about how they made the serum.

'We start with this.' Professor Stravinsky bent down and took a small bottle, marked **AHS-V**, out of the fridge. 'This is the virus, in some infected tissue.'

He went on to explain how they neutralised the virus and injected it into the Kladruby horse Tomas had just shown me. The horse's immune system then took over and began making the antibodies which would fight the infection. It was a complicated process, taking about a month before the serum was made. Then it would be ready to inject into any sick horse that needed help in fighting the virus.

Husak took a small bottle from the fridge. 'Here we have the start of the serum. It is still early days, but very soon no horse need die of African Horse Sickness.'

We had tea in what had been a grand drawing room before we said goodbye. I was very impressed with Tomas's father and I had to admit, I was getting more and more impressed with Tomas as well!

That night, after dinner, we all felt a great bond developing between us. We were united in our determination to win this great race and we raised our glasses to salute the horse we hoped could do it.

'To Paddy, a truly great Irishman!' proposed Ben.

Edward held up his hand and cautioned us. 'Much as I prefer British-bred horses, on this occasion I will join you in a toast to an Irishman! To Paddy!' said Edward.

'To Paddy!' we chorused, our response echoing into the night.

CHAPTER SEVEN

Race day dawned bright and clear, but I shivered momentarily as I closed the lorry door quietly behind me and slipped across the grassy compound which separated the trailer park from the stables. There'd been a heavy dew overnight and I tested the ground as I made my way across. Good! The going would be perfect. The stables were beginning to stir and I smiled at one or two of the other lads who'd decided to get up early, too.

We'd walked the course the day before, and although we'd read about the Pardubice and watched videos of some of the old races, nothing could have prepared us for the reality of what we saw. The sheer size and scope of the fences had been overwhelming and we'd felt dwarfed by them.

The notorious Taxis fence had lived up to its reputation. It was by far the biggest on the course and it was only jumped once every year – now I knew why! The natural hedge was massive, and there was a yawning ditch and a huge drop on the other side. The spread was enormous. Paddy would go for this fence – it was just the sort of challenge he liked.

The big danger was from loose horses that could interfere with Paddy's stride; this was the fence that mattered and we all knew that if Paddy cleared it, he'd have a real chance of winning. The course was over four miles long and we'd returned tired but excited, our tactics decided for the race the next day.

'Agreed!' Edward had summed up his strategy. 'I'm going to stay on the outside rail and run fifth or sixth, that way there's a chance I'll stay out of trouble.'

Later, Edward was in one of his rare good moods, as I'd given him a leg up for that afternoon's schooling session over the course. He'd wanted to know what I'd been able to find out about the other horses and their jockeys. Of course, I'd been studying the opposition for all I was worth!

'Tomas isn't here just now, but his horse, Genesis, looks a goer, and that one over there,' I'd pointed to a small thoroughbred, 'that's another good Czech horse called Turandot.'

'Well, I don't think these Eastern rabbits are going to give us much of a race!' Edward had snorted, polite as ever!

I must admit I'd been surprised when I'd seen how different their thoroughbreds were from ours. The Czech horses were tiny and didn't seem to have much substance at all.

'And their jockeys aren't much bigger than rabbits, either!' said Edward, adding insult to injury as he pointed to one of the smallest Czech jockeys.

'He's Dominic Vitek,' I'd said, quickly. Edward was so rude! 'Not just any old "rabbit", you know! His father's a top trainer and his grandfather won the Grand Pardubice twice! But the one to look out for,' I'd added, lowering my voice, 'is a Russian-bred horse.

Look, there he is!' I'd pointed at the huge black thoroughbred, Romanov the Third. 'Isn't he something?'

The practice session had gone well enough, but when Edward had returned to the paddock he was very angry.

'Where's Ben?' he'd shouted at me. 'These so-called jockeys intend to stitch me up! I was boxed-in out there and Paddy didn't like it, either! What are we going to do about it? Sportsmanship clearly doesn't count for much in this wretched country!' Edward jumped down from Paddy and threw me the reins.

Very few foreigners competed in the Pardubice and Edward was sure the local jockeys had it in for him.

I'd bittten my lip, furious that Edward was being so rude again. It was true, I'd heard these rumours, and I'd told Ben about them but we both thought it was nothing more than talk – and we'd decided not to pay any attention to it or bother Edward with stable gossip.

Ben had come over when he heard the noise and had tried to calm Edward down. But without much success. Edward had seemed horrified by the prospect of two Czech jockeys planning to interfere with his race by riding on either side of him and effectively stopping him and Paddy from being able to move where they wanted. It was an illegal and dangerous practice.

Ben had tried to convince him that the locals had just been trying to freak him out, but Edward had only calmed down when we'd suggested a change of tactics.

'Paddy has an enormous gallop and we're not worried about him running out of energy,' Ben had reassured him. 'Forget about hanging back – tuck yourself in behind the front runner, who'll probably be clear on the inside rails. If you're up front, no-one can box you in!'

So it had been decided. But I'd felt uneasy as I'd settled Paddy down for the night. Had I got my training programme right? Would Paddy start the race this time? Edward would be riding without a whip, so we'd soon find out. But now the big worry was whether Edward would keep his cool. I tried to convince myself he would, but my stomach had begun to churn in a very nasty way!

The Grand Pardubice was a big occasion in the Czech calendar, and the spectators were arriving in their thousands. There was a carnival atmosphere at the course with many stalls to buy from and gypsies and pick-pockets, too! Soon the grandstand was full and the huge crowd was buzzing.

I'd made a special effort and secretly I thought that Paddy looked wonderful when I led him into the paddock. But my heart sank when I saw Edward striding across in his familiar purple silks with yellow spots. He looked nervous and agitated and I felt uneasy as Ben legged him up into the saddle. I headed for the stand where I said a silent prayer. There was nothing more I could do. The race was called and now it was up to Paddy and Edward!

I hardly dared breathe as I watched the horses canter to the starting line. This was our first test and Edward took Paddy well up the field to avoid the worst of the crush. Suddenly, there was a disturbance. I re-focused my glasses and sighed with relief. It was Turandot throwing a wobbly – not Paddy! He was quietly edging his way back to the tape as the starter put them under orders. My heart was beating fast - they were off!

They all cleared the first and as they came past the grandstand Paddy was looking good, eating up the ground with his huge gallop, as Edward worked him through the field. The Russian horse was in the lead and now Tomas was making ground. They'd soon be at the Taxis and I could hardly bear to watch! Several horses were very close and I could see Edward was tense.

'Go on, Paddy!' I yelled. 'Go for it!'

Paddy could see his way ahead and took off as two of the horses ahead of him fell – one of them badly. But Paddy sailed right over, somehow avoiding the pile-up.

'Well done! That's my boy, Titch!' I breathed a sigh of relief. The worst one was over! I began to cheer, and then momentarily froze. Now the other jockeys seemed uncomfortably close and I could see that Edward thought he was being squeezed. Twenty paces away from the next jump, Edward still had a rider on either side of him. Ten paces away and they were still there, and now he looked dangerously uncomfortable. I was sure Paddy could easily have accelerated away from the rest, but as the fence drew near Edward panicked, missed his timing and got it all wrong. Paddy took off far too close to the jump, ploughing through it, clearly unhinged.

From there on, it was as good as over. The rest of the race passed in sickeningly slow motion and it must have been a terrifying ride for Edward, as he struggled to control both the horse and his panic. Our careful race plan had been abandoned, and now Edward just struggled to get home in one piece.

It should have been one of the proudest moments of my life, but it was a disaster.

Instead of crossing the line first, Paddy had come in twelfth, well down the field. I shuddered, but my deep, muffled cry was drowned out by the roar of the crowd, pushing and cheering around me.

I tossed back my hair and brushed away a tear as I stumbled back through all the people, only vaguely aware that Romanov had crossed the line first. And I was sure Tomas was up there with the winners, too.

I tried to calm down and hide my disappointment before meeting Ben, who'd be leading Paddy round the back towards the stable block.

When I saw him, Ben squeezed my arm as I took over the leading rein and he quietly mouthed, 'Bad Luck, Becky'.

I smiled, weakly, grateful for his support, and turned to Edward who was looking grim. 'I told you they were going to stitch me up!' he snorted. 'It's just the sort of behaviour you'd expect in a country like this! Half of them are gypsies!' Edward was out of order.

'So much for you trainers and your pseudopsychology! Just trying to freak me out with talk, were they?' Edward was sneering, now. 'They stitched me up! I told you they would!'

'Let's get Paddy back in his stable,' interrupted Ben, taking control, 'and then we can have a post-mortem when we're all feeling a bit calmer.'

'You can forget your post-mortems!' yelled Edward. 'I want a stewards' inquiry, and I want it now!'

I bit my tongue and handed the saddle to Ben. He set off after Edward, who was angrily striding away. It was true, there'd been a bit of a jostle, but I didn't think it'd been anything serious. In my opinion, Edward had panicked. Paddy's breathing was slowly returning to normal.

'It can't have been great for you either, Titch!' I said under my breath. 'Oh how I wish I'd been riding you! *I* wouldn't have been frightened. We'd have just gone for it!'

But it wasn't meant to be. I sighed as I got on with the long job of bedding him down and putting everything away. The good luck cards on the door now looked horribly out of place and I tore them down. There was one from Claudia, and one from Sue and the team, of course. But the one I was hoping for was missing. I hadn't heard a word from Jamie in weeks! I put my head on Paddy's neck and cried.

Ben was looking grim when I eventually made it back to the lorry.

'The stewards denied the inquiry, of course, but Edward insisted we demand one. It was quite clear – we watched the re-run and there was no interference. Edward's had to accept it.'

There was a knock on the door and Tomas appeared, bright-eyed and exhilarated.

'Congratulations, you star!' We all tried to be cheerful. Tomas was in the prize money; he'd come in third.

After we'd hugged him he announced some news of his own in his strangely formal way.

'I come from my father to ask you to eat dinner at our house. We have some good news for celebrating. My government have money for sport and now they give it to me to go to

England to train!' Tomas lowered his eyes, took a deep breath, and continued. 'My father ask if you think you can take Genesis and me in your stable. Please, talk about this after dinner.'

My heart missed a beat and I felt the colour rise in my face. Tomas was welcome in my stable any day of the week! So he was coming to England and probably to the Mainwaring's. That meant I would be seeing a lot more of him from now on!

CHAPTER EIGHT

As we turned into the familiar lane leading to the Mainwaring's yard, I felt a huge sense of relief to be back home with the nightmare of Pardubice behind us. Ned and I had agonised for hours on the journey back, analysing the race in detail, trying to assess what had gone wrong. But the only conclusion we could come to was that it had to be something to do with the jockey.

Edward was an amateur, a talented one, but he had a lot of business needs to attend to, and his riding often took second place. He'd been under a lot of pressure before the race and hadn't been able to make all his training sessions. He hadn't managed to familiarise himself with Paddy as much as he should, and the sheer size of the fences at the Pardubice would have filled any jockey with awe. Whatever the reasons, there was no doubt that Edward had panicked.

I almost felt tears in my eyes as Scruffy came tearing round the side of the house to greet us. 'I've missed you too, you stupid pooch!' I laughed as he launched himself into my arms. 'You can't fool me with all those licks! I bet you were up to no good when we were away!'

'Darned right, he was!' laughed Claudia as she enveloped Ned and me in her familiar scent-filled embrace. 'He's the talk of the village! He jumped out of the car window when I parked at the hairdressers, and you should have seen his performance with Mrs Pearson-Craven's poodle when I caught up with him! Well, it left nothing to the imagination, my dears! The faces of the old ladies under their hairdryers were something else. Overheated isn't the word!'

Jessie, John, Patsy and Andy were all

there to help de-rig the lorry and get Paddy settled back home. As soon as I could, I shot off to see Monkey who was standing quietly in his stable. He signalled his pleasure at seeing me with the familiar movement of his ears.

'Oh, Monkey! It's great to see you again. We've had a miserable time.'

I jumped as Patsy appeared round the door. 'Despite everything, Mr Holsborough's just phoned to say he's sending over the two youngsters tomorrow. Do you want me to get John to deal with the arrangements?' Patsy could see I was upset.

'That'd be great, thanks Patsy. I'll be quite busy washing down the new stable ready for Genesis, who's coming tomorrow.'

'Make sure you leave time to wash your hair,' joked Patsy, quickly. 'I hear this hunky guy Tomas is coming with Genesis, too!'

I knew I was blushing as I dismissed Patsy with a casual toss of my head. Then I smiled to myself. Perhaps I would wash my hair after all!

Tomas quickly settled into the routine at the yard. He stayed in the hostel with the other lads and was eager to learn all he could. He soon became very popular with everyone as he mucked in (and mucked out!) wherever he was needed.

At last I'd heard from Jamie. He'd phoned just after we got back, to try and cheer me up. Apparently he had sent me a good luck card, but it arrived after we'd left. I'd been really disappointed when he'd said he wouldn't be around for Christmas, but he'd promised he would see me in the New Year.

In any event, Christmas Day had passed without much fuss. It wasn't a big deal for those of us who were working. The horses still had to be fed and there'd been no let up in training – three of our horses had run in the King George V Stakes at Kempton Park on Boxing Day. Red Rag was just beaten into second place and Genesis did surprisingly well to come in fifth, well ahead of Mr Turner's Millennium Mount. We were in good form on the drive back.

'That woman from the *Racing Post* was very keen to interview you, Tomas! I think she fancies you!' I ducked as Tomas threw an empty drink carton at me, but I hadn't finished.

'You're becoming quite a star, you know! You'll be on TV next!' I liked to tease the poor boy!

'I will tell them all about my excellent trainer, who rescued me from the Czech Republic and help to make me so good.' Tomas gave me a big smile and I started to blush. 'Yes, I will tell them all about Mr Mainwaring.'

Now it was *my* turn to throw something at *him!*

'Hey, you two! Simmer down, will you? That nearly hit me!' Ben growled at us. He screwed up his eyes and peered through the drizzle at the deserted road ahead. 'All I want is to get home in one piece and put my feet up. I don't know where you get the energy!'

'From winning – or nearly winning!' I was serious now. 'Monkey was brilliant! If only Johnny could have coaxed a little bit more out of him.'

Johnny Sutherland was one of the best steeplechase jockeys around and it was a bit mean of me to put the finger on him.

'Aha!' Ben had perked up a bit now. 'D'you hear that, Tomas? She's blaming the jockey now. The next thing you know, she'll be wanting to ride him herself!'

We laughed, but the Governor wasn't far off the mark. I would never be a professional jockey, but after riding horses like Monkey and Paddy, I would certainly be interested in being an amateur one. I changed the subject quickly.

'One jockey we haven't seen much of recently is Edward. Have you got any news of him?'

Edward had decided to rest Paddy after the Pardubice. He wanted him to race at the end of the season, at Cheltenham, and of course at Aintree. The Grand National was the one he wanted.

'Apparently he's been very busy developing some new business interest. He'll start again in a month or so. Until then, Paddy's in your capable hands.'

'I bet he didn't say that! The day Edward gives me a compliment, I'll drop dead from shock!'

Strangely enough, the next time I saw Edward, he did give me a bit of a shock!

It was the middle of February and Edward had just come back to work. He said he had a business proposition for me!

'I imagine they pay you peanuts here,' he said, eyeing the premises. 'How would you like a bit of freelance work on the side, to earn some extra cash?'

He produced a smart navy-blue leather briefcase embossed with gold lettering. 'And you needn't worry, I've run it past the Governor, as you so charmingly call him, and he's given his approval.'

He snapped open the briefcase and inside was a tester spray and several presentation bottles of *Charisma* – a new scent which Edward wanted to promote on the up-market racing circuit. He sprayed some on to my wrist.

'Don't you think it's got a certain something?' he asked, as I sniffed it appreciatively.

Amongst his many business interests, Edward had recently taken over a cosmetics company and he quickly sketched out what the deal would be. I could use my contacts in the racing world – owners' and trainers' wives and girlfriends – and for every bottle I sold, I would get twenty per cent commission, cash in hand, and a free sample for myself!

'It sounds good!' I hesitated for a minute, but I couldn't see any catches. And I had to admit I could do with some extra money. Ben's wages didn't stretch to much, and I was beginning to take a serious interest in clothes and CDs.

'It's a deal!' We shook hands. 'In fact I can

start this week. Red Rag's injured his fetlock joint and the Governor has arranged for him to go over to the Knight's yard for some swimming therapy. I'll be driving him over on Friday, so how's that for business?' I exclaimed.

'Well, it's a start,' said Edward, and he handed me the briefcase. 'I expect you to be business-like and send me the invoices so I can keep a check on your sales. We have to keep the tax man happy, you know.' He laughed nervously, handing me a small cardboard box. 'And here are a few samples for the grooms. You can distribute them as you see fit.'

Edward had certainly mellowed since I saw him last. He'd never thought of the stable girls before. Maybe he'd undergone a personality change!

'And you needn't worry about the ones here. I've already given Patsy and Jessie some, too. Well then, here's to *Charisma* and success!'

'Keep your head down!' warned Jessie, as I prepared to get Red Rag ready for his move to the Knight's yard. 'Patsy's in a really bad mood today!'

'What's new?' I replied with a grin. Patsy wasn't the easiest of people.

'She's furious that Tomas is going with you this afternoon instead of mucking out like the rest of us mortals! As an owner's daughter, *she* would have liked to visit one of the big racing yards!'

'Thanks for the warning, Jessie! Life's tough at the top! By the way, let's go shopping on our next afternoon off. I could do with some advice on a new dress I want to buy.'

I liked Jessie a lot. After Chrissie, who was studying journalism at a college in London, Jessie was my best friend – and now Tomas was a friend, too. Suddenly I thought of Jamie, far away in Ireland, and I felt guilty. I reminded myself that I'd nothing to feel guilty about, and in any case, Jamie and I had agreed that we should both have other friends.

The sun was shining as Scruffy and I joined

Tomas in the Landrover ready to tow the trailer over to the Knight's. I carefully laid the *Charisma* briefcase on the back seat and told Tomas I hoped to do some of my freelance sales work while we were there. He was very impressed! He was even more impressed when, after half an hour's drive through the quiet country lanes, we reached the smart Lambourne yard.

'Wow, this is beautiful!' he gasped, as we swept through the imposing main gates and up the manicured gravel drive towards the stable block. The place oozed money. The Knights had impressive clients and I secretly hoped we might run into one of the rock stars who had horses in training there.

'Now I see why you wear new clothes to come here!' Tomas ran his hands through his hair and pulled at his jacket.

I laughed, but it was true. I'd made a special effort with my appearance for my first sales trip. I'd brushed my hair back and chosen a green top which went well with the tawny colour of my hair. I was wearing my best stretch trousers and a figure-hugging jacket. Secretly, I thought I looked really good! I'd arranged to meet Kirsten Knight, the trainer's glamorous wife, and she'd invited one of her friends along, too.

But first, Red Rag was to get his swim and Greg, their head lad, showed us how to do it. The swimming pool was great. It was circular, and I was to walk round the side, holding a long wooden pole which hooked on to Monkey's halter, to keep his head out of the water.

I was expecting trouble as I gently led Monkey down the ramp. To my knowledge he'd never swum before. But I needn't have worried, he took to the Knight's swimming pool like a professional. Greg said he'd get about five minutes every day and after a couple of weeks he should be fit again.

Kirsten and her friend joined us in the stable, just as we were finishing with Red Rag.

'So this is the famous Gold Cup Winner! We're very proud to have him staying here for a couple of weeks. Our rock star owners will be very impressed, too,' she laughed. 'This is Julia Campion, who's always looking for a new scent!' The two friends began to giggle. They were both really attractive and suddenly I felt very ordinary.

'We're going out for lunch in a minute, so to save time, could you do your sales pitch out here? Thanks, Becky.' Kirsten flashed her perfect white teeth in a dazzling smile as I ran to the Landrover to get the briefcase.

'It's a big improvement on the normal smell in our stables,' grinned Kirsten, as she sprayed the sample on her friend and examined the bottle, 'and it looks good, too, doesn't it? I love having beautiful things on my dressing table.'

'I'd rather have something beautiful on my arm,' grinned Julia, 'and I think this'll do the trick! I'll take one, thank you.'

And so my first sales were made – one bottle to each of them. Hurriedly, they scribbled me a cheque and wafted out of the stables to get on with the important matter of lunch.

'Phew!' I gasped, wrinkling my nose at the

overpowering smell which now enveloped the stable. Red Rag swished his tail. 'I don't think Monkey likes it, but it's the fastest money I've made in a long while!'

I could never have known it at the time, but it was also the most expensive mistake I'd ever make, too!

CHAPTER NINE

'It's not fair! You look gorgeous!' moaned Jessie, as she admired the short black dress we'd found in a new shop in Newbury. 'Some people have all the luck – a fabulous figure, lots of money *and* no spots!'

'I don't know about the fabulous figure,' I said as I stood sideways to the mirror and pulled in my tummy, 'but I'm really glad of the extra money I've earned from the *Charisma* agency, otherwise I wouldn't be able to even *think* about buying this.' I slowly fingered the price ticket. Sales had gone well and I'd had plenty of opportunity to push the scent as hard as I could.

'Do you think Jamie will like the dress?' I twirled around in front of Jessie.

'He'd be mad not to,' she sighed. 'Go for it, Becky! It's perfect!'

'I'll need new shoes... ' I hesitated, 'but yes, you're right, it's great and I just have to get it.'

I was very happy. Jamie had written to say he was coming over for a long weekend and was going to the Doncaster Horse Sales, as a guest of one of the Irish bloodstock agents. It would be a mixture of business and pleasure, and he'd wondered if I could come along for the pleasure bit! I might enjoy the horses, too, he added! There would be a dinner party and a disco after the first day of the sale, and he knew of a small hotel nearby where we could stay – separate rooms, of course, and he hoped Ned and Ben would allow me to go.

Ned wasn't too keen, but Claudia went to work on him and reminded him I was seventeen, (OK, only just!), and said she was sure Jamie was responsible and they'd have to trust me! In any case, Jamie wouldn't have time to see me otherwise, and it would be spring before he was back in England. So, like Cinderella, to the ball I would go – and now I had the perfect dress!

Jamie had arranged to pick me up mid-morning at the Mainwaring's yard. He wanted to see our new star, and all the improvements Ben and Sue had made since he rode Red Rag at the Cheltenham Gold Cup. They were just coming back from looking at the new stables when I drew up outside the house. Even though I was expecting him, I felt my face go hot when Jamie smiled and gave a low whistle. 'I see the improvements haven't been restricted to the yard, Pusscat!' He brushed his lips across my cheek and added, 'Mmm, and you smell good, too!'

I'd sprayed my hair and wrists liberally with *Charisma* and I said a secret thank you to Claudia, who'd given me the beautifully-cut jacket I was wearing.

Scruffy jumped up between us, almost jealous, and suddenly I didn't feel embarrassed any more.

'I don't believe you've met before. Scruffy, may I present Jamie Howland. Jamie, this is Scruffy, my dog!'

'Don't tell me,' laughed Jamie, anticipating my next request, 'you're inseparable and you'd like him to come with us! Well, only if he spends the night on a rug in the back of the car.'

'Thanks, Jamie.' I gave him a quick hug and was about to get into his bright red, shiny car when Sue came running out of the house, white as a sheet, and waving a letter in her hand.

'Ben, it's awful news – dreadful – I can't believe it's true!' She hardly paused for breath as she thrust the letter into Ben's hand. 'Oh Ben, if it *is* true this'll be the end of the yard!'

We followed Sue back into the kitchen and Ben studied the letter, the colour draining from his face, too. Finally, he read it out aloud.

> *'We are in possession of a lethal virus and we intend to use it in the top steeplechase racing yards. Your horses will be infected, and will probably die unless each yard pays two hundred thousand pounds into a numbered Swiss Bank Account within the next seven days. On receipt of the money, a serum will be delivered to you, which will cure*

the infected horses and protect the rest of your stables from this deadly disease.'

Ben looked up and there was a hollow ring to his voice as he added grimly, 'The letter goes on to give the address and number of the Swiss bank account.'

'It's got to be a joke,' I said. 'It sounds just like the type of thing Professor Stravinsky was working on. I'll go and ask Tomas if he knows anything about this.'

Before I could move, the phone rang. It was the Knights and they'd just received a letter, too, and several other trainers had been on the phone about ransom demands. Suddenly, it looked very serious and I ran over to the stable block to get Tomas to come and phone his father.

The news from the Czech Republic was bad. Husak had vanished ten days before and the serum was missing. The Professor had thought it was an internal matter and the police were trying to find Husak to charge him with industrial espionage. But after Tomas told the Professor about the ransom letters, he knew it was more serious and a quick check of the laboratory confirmed everyone's worst fears. The jar with the virus had vanished, too. The threats were real! Husak had the means to infect the horses. Finally, and Tomas was wringing his hands as he told us, 'My father is very distressed to hear your news.'

'Distressed!' yelled Ben. 'Not half as distressed as I am to hear that his assistant is roaming the country, and possibly my yard, with that lethal stuff in his pocket! I'm getting on to the police immediately!'

'Well, at least there's no sign of it here yet,' said Sue, trying to calm things down. 'All the horses are well and I haven't seen any strangers round here. If we step up security and the police get on to it straight away, we should be OK, surely?' she added desperately. 'There's no way we could find two hundred thousand pounds.'

I turned to Jamie, about to say that I thought we should cancel the weekend, but Sue saw my anxious look and quickly intervened.

'Becky, there's no point you staying here. The police will take care of things and hopefully, by the time you get back from Doncaster, they'll have found that awful man.' She looked at Tomas. 'I'm sorry, Tomas, I hope he isn't a friend of yours.'

Ben gave Tomas a comforting pat on the back. 'He couldn't be, not after what he's done to your father,' he said, standing up and making a big effort to sound cheerful. 'Now off you go, Becky!' He wagged his finger at Jamie, teasingly. 'And as for you, young man, you'd better remember that Becky's my number one trainer, so look after her, or you'll have me to answer to!'

'Number one trainer?' I quizzed him. 'Can I quote you on that?' I tried not to look too embarrassed as we got up to leave. 'Seriously, Ben, I'll come back at a second's notice, if you need me. I hope the police sort it out soon.'

I put Scruffy on the back seat of the car, beside a basket with a thermos of coffee and cream, and slammed the door of the MG.

Jamie swung the car down the drive and we were off for our precious weekend.

The ransom letters were the talk of Doncaster. Most of the trainers had received identical threats and they were taking them really seriously. Like Ben, they'd decided to step up security at their yards, and hoped that this would work. I felt very nervous myself.

The Doncaster sale was one of the top auction sales for thoroughbreds, and I'd never seen so many horses in one place. There were literally hundreds of them – all listed and described in the bulky catalogue. It was very exciting.

I fell in love with a fabulous black thoroughbred but unfortunately the prices were as fabulous as the horses. It would take a big Lottery win before I could look at any of them. I hunted for Jamie to see what he thought of the black horse, but he'd vanished – again! He'd spent most of the day socialising;

networking with the owners and trainers, making as many new contacts as he could. I suppose he had to do it, but I was getting more and more fed up being left on my own. Eventually, I found him talking to some boring-looking man.

'Jamie, do we need to stay much longer?' I whispered, when they'd finished talking to each other. 'We'll need to get changed for dinner soon.'

Jamie frowned as he looked at his watch. 'Mmm, it's later than I thought. We'd better get back.'

At the dinner-dance Jamie wasn't much better. Although he'd managed a low 'Wow!' when he'd seen my dress, there was still an endless round of socialising to do and a lot of it involved Jamie chatting to the owners' daughters. Naturally, he found that part pretty easy. I felt close to tears as the lights dimmed and he led me on to the dance floor towards the end of the evening. There was only half an hour of the dance left, and I felt cheated – he'd hardly been with me at all.

Later, as we drove back to our small hotel, I felt miserable. I said goodnight and went up to my room as soon as we got back. I felt tired and confused. Maybe I'd read more into the birthday card and the weekend's invitation than there was meant to be. And what about that newspaper photograph of Jamie with Sophie Redmond? I threw myself on to the bed, tears threatening to overwhelm me, when there was a quiet knock at the door. I heard Jamie's voice.

'Becky, are you awake?'

I tossed my hair back and straightened my dress, smiling bravely as I let him in. He looked wonderful, standing in the doorway, his dark hair falling over his eyes and a concerned expression on his face.

He pulled a small package out of his jacket pocket and gave it to me, hesitantly.

'Becky, I wanted to give you this, to thank you for putting up with me today. I can't think of anyone else who would have stayed so cheerful, with all that business I had to do.'

I could feel myself blushing as I took the

beautifully-wrapped present.

'But,' he went on, 'I wasn't sure whether you'd want this or not – you seemed rather distant this evening.'

I tried to interrupt to explain but he held a finger to my lips. 'Let me finish, Becky. I've been away a long time now and I couldn't help noticing how friendly you are with that Czech boy, Tomas. I know we said we wouldn't get too serious at this stage and—' he paused for a moment and then looked sad, 'perhaps you prefer him to me.'

'Oh, Jamie!' I threw my arms round his neck. 'You've got it all wrong.'

He let out a big sigh and kissed my neck as I burbled on. 'Tomas is just a friend. He doesn't know anyone else in England and he's latched on to me, but it doesn't mean anything! I thought *you* didn't like *me* any more! There was a photo of you in the paper with Sophie Redmond and I thought... ' My voice trailed off as Jamie laughed.

'I hope I've got better taste than that! No, Sophie just saw a good photo opportunity and gatecrashed the picture! She's going out with one of the jockeys over there.'

Remembering the present, I tore open the wrapping paper and gasped as I saw a gold chain with a pendant in the shape of a letter B.

'Oh, Jamie!' I gave him a big kiss.

'So you like it?' There was a touch of anxiety in his voice.

'Like it? You idiot, I love it!' And I bit my lip as I nearly added, 'and I love you too.' But I held back – it was still early days.

'It's B for Becky and B for Bossy!' Jamie was smiling, cheekily. Before I could hit him he pulled back from my arms and said, 'You smell far too good for your own safety, Pusscat. Anyway, I promised Ned, Claudia, the Governor and almost the entire world that I'd take care of you! So I'll say goodnight!'

I slept on-and-off the whole night, my emotions in a delicious turmoil, and I felt a little unsure of how to greet Jamie when I went down to breakfast the next morning. I needn't have worried – Jamie was already out at the car. He'd taken Scruffy for an early morning walk and I was pleased to hear he'd behaved himself overnight in Jamie's posh car.

The atmosphere was only slightly strained when we started our drive back home. 'Sit down, Scruffy!' I urged, as my dog tried his usual trick with his paws on the back of my seat; not an easy position to find in a sleek and shiny sports car! Jamie put his foot down and swung the car round a corner. 'Oh, no! You stupid dog!' I shouted, as Scruffy toppled over, spilling the remains of yesterday's cream on to the back seat. Furiously, I tried to mop up the worst of it with a handful of tissues, but in no time at all, the car began to stink!

Jamie and I burst out laughing, any tension forgotten as we continued our journey. When we finally drew up at Nettleton Lodge, Jamie told me he'd definitely be back for the Cheltenham Three Day Festival in March, and he'd see me then. I ran inside, feeling really happy, to give Ned and Claudia all the news. But one look at their faces told me something dreadful had happened while I'd been away.

CHAPTER TEN

'It's Red Rag, Becky,' said Ned baldly, putting his arm round my shoulder and guiding me towards a chair. 'We tried to ring you at the hotel this morning, as soon as we heard, but you'd already left.'

'What's the matter with Monkey!' I cried, and already a dreadful feeling was rising from the pit of my stomach.

Ned spoke quietly. 'I'm very sorry, Becky, but he's gone down with a horrible lung infection and I'm afraid it looks as though it's the virus.'

'I'm going straight over to see him,' I said, snatching up my coat. But Ned pulled me back on to the chair.

'It's not as simple as that, Becky. Monkey's

in quarantine now and the yard's been sealed off. The Knights have said no one can come and go any more. Why don't you go over to Ben's straight away and see what's to be done?'

As I ran round to the garage where my car was parked, tears were coursing down my face. I really don't remember my drive to the yard, but when I drew up at the Mainwaring's, a police car was parked outside the kitchen and Ben and Sue were saying goodbye to one of the officers. I took a deep breath before I dared to ask. 'How is he, Ben? How's poor Monkey?'

'The news isn't good, I'm afraid, Becky.' Ben was as white as a sheet. 'The Ministry are running urgent veterinary checks, and we're still waiting for the results. We've rung the Czech Republic again and it looks as though the symptoms, so far, are exactly the same as those caused by the missing virus.'

'What about the rest of the horses?' I asked quickly. 'Are they ill too?'

'So far it's only Monkey, but there could be others at any time.'

Patsy was coming out of the house, carrying a notebook. 'Governor, that was the Knights again. Two of their horses have gone down with the virus and Kirsten Knight said to tell you they've just heard from the Sullivans – they think one of theirs has the illness, too.' Patsy flicked through the notebook. 'None of them have done anything about getting the ransom money together because the police said it was probably a hoax and not to pay.' Patsy stabbed at the page. 'I've made a list of all your owners, Governor. Shall I start to ring them?'

I hardly dared to ask the next question. 'What are Monkey's chances, Ben, if he does have the virus?'

Ben lowered his eyes and looked at the gravel. 'We haven't got any serum and without it I'm afraid he could die.'

I turned blindly into the yard and flung myself into the first empty stable I could find. Scruffy ran after me, puzzled by what was going on, and he whined and licked away the salty tears as they streamed down my cheeks and fell into the straw. It was half an hour before I felt I could return to the kitchen, and my eyes were stinging and red.

'What are we going to *do*, Ben?' I gasped, urgently. 'Can't you get the serum?'

'We're phoning all the owners right now but I'm sorry Becky, it's going to take several days to raise the finance. You know I'd do anything to help Red Rag but two hundred thousand pounds is out of my league.' Ben bowed his head as I searched for something positive we could do.

'Can I go and stay at the Knight's, to be with him?'

'I've already thought of that, Becky, but whoever stays with him also goes into quarantine – and we really need you here.'

My face fell as Ben went on. 'Infected yards won't be able to race but for the rest of us, the racing season will carry on and Paddy's at a crucial stage of his training.'

He turned to Sue.'I think we should send Jessie over to stay with Red Rag. She's the best stable-hand I've got and she has an intuitive grasp of illness and what makes horses tick. If anyone can pull him round, it'll be Jessie.'

I was disappointed but I had to agree. Jessie had some special qualities: she was almost a healer, and her quick diagnoses had often made a big difference to a horse's recovery.

'She gets my vote, too,' I added quickly. 'Shall I get her Ben? You'll want to brief her as soon as possible, won't you?'

'That's my girl, Becky!' Ben smiled weakly and then frowned. 'But it's strange the way Monkey was the first to go down with this, and it's even stranger that there's no sign of the illness here.'

You could have cut the atmosphere in the yard with a knife, and even the pop music we always had on began to sound eerie. We all seemed to be on automatic pilot and somehow staggered through the day like zombies. We waved Jessie off with a tearful goodbye; she'd gathered together everything she could think of that might make a difference to Monkey – she even took his special Gold Cup presentation rug, in case he needed the extra warmth.

By the end of that day the news was even worse. The Ministry confirmed it was the same virus which had swept through Australia. In the week since they'd received the blackmail letters, the trainers had gone from disbelief to red alert. They realised now that the extortionists meant business, and even though police advice was *never* to pay blackmailers, the trainers were busy contacting all the owners for their share of the ransom money. Life was awful!

It was getting dark and Ben's headlights cast an eerie light over the yard as he drove Jessie over to the Knight's. The shadows seemed menacing and all the Mainwaring staff were jumpy as we kept a watch over our horses, looking for the first signs of the virus. It was the worst day of my life and I went home to Ned and Claudia feeling really low.

'Have the police got any new leads?' asked Ned putting his arm round me. 'You look ghastly, Becky.'

'I'm fine, Ned, don't worry.' I looked up at him sadly. 'At least I'm a lot better than Monkey is right now.'

'What's the latest news on Monkey?' Claudia appeared from the sitting room shaking her head. 'It's like a nightmare. I can't believe this is happening.'

'Monkey's very ill and there doesn't seem to be anything anyone can do.' I could feel the tears welling up behind my eyes and I mumbled some excuse and went up to my room.

I didn't have too long to feel miserable before Claudia yelled up the stairs. 'Telephone, Becky. It's Chrissie.'

It was like talking to someone from another world. Chrissie had no idea what was happening, and for the first five minutes she burbled on about London – the work-experience job her college had found her as researcher on a teen magazine, and a new bloke she'd just met.

'But I don't know why I'm telling you all

this now, Becky,' she went on. 'The real reason I phoned was to see if I could come and stay with you this weekend.' She paused a moment and then asked. 'Is there something wrong, Becky? You seem very quiet.'

It all poured out then and I told Chrissie everything, just about sobbing by the end. 'So you see, Chrissie, it's impossible. I wouldn't be much fun at the moment.'

When I went back into the sitting room, Ned and Claudia said they couldn't help overhearing some of the call and they thought it would be a good idea if Chrissie *did* come for the weekend. It took them a minute to persuade me, but eventually I had to agree that it would be great to see her. I rang Chrissie back and said it would be OK after all.

I'd just emerged from a long soak in the bath when Claudia shouted up the stairs again, this time with a lightness in her voice.

'Becky, you're a popular girl tonight. There's another call for you – this time it's from Ireland!'

I shot down the stairs still wrapped in my

towel, and grabbed the phone from her hand. 'Jamie! It's you! How are you?' The words came out in a breathless jumble.

'I'm OK, Pusscat.' I smiled as I heard his pet name for me. 'But more to the point, how are you? We've just heard that the virus has been confirmed.' He took a deep breath and went on. 'Is it true that Monkey's got it?'

I really poured out my heart this time, in a way I couldn't to anyone else but Jamie. Red Rag was special to him, too, and he understood the bond between me and Monkey. But Jamie was practical and he started firing loads of questions at me.

'How bad is he, Becky? Is he coughing a lot? How high is his temperature?'

I explained that I hadn't seen him myself but Jessie had telephoned us. 'Jessie says he's very weak; he's not eating and he's very listless. She says his cough is awful.'

I started to cry. 'Oh, Jamie, where's the magic serum? Surely Monkey can't die?'

'I know they're doing their best to raise the ransom money but it all takes time.' Jamie sounded tired.

'But time is something Monkey doesn't have!' I could hear my voice rising with panic.

Jamie thought for a moment. 'Look, it doesn't sound too good, Becky, and you'll have to prepare yourself for the worst. The vets say the next three days will be crucial. If his temperature goes on rising, his heart won't be able to take it.'

'Do you mean if he *does* get through the next three days, then he might stand a chance?' I was ready to grasp at any straw.

Jamie paused. 'Red Rag was in pretty good shape when he went into this illness, in fact, in a way, his joint injury couldn't have been better. All that swimming can only have strengthened his lungs.' Jamie's voice wavered a little. 'I'm sure he'll put up a big fight.'

There wasn't much else to talk about but, before he went, Jamie made me promise to phone him if there was any change at all. I felt drained and I crawled up to bed for an early night. As I settled down to sleep, I felt Scruffy making himself comfortable at the foot of my bed, snuggling down into the duvet near my feet.

'OK, Scruffy, your luck's in. I could do with a friend tonight.' I sighed as I turned out the light.

Edward Holsborough's familiar Mercedes was parked outside the kitchen as I drew up in the yard the next day. He hardly dared to ask how things were but he was relieved when we told him that Paddy and his two young horses were OK. He looked shocked to learn that Red Rag was so dreadfully ill.

'So sorry to hear the news, Becky.' He squeezed my arm and muttered distractedly. 'It's a horrible world sometimes – a horrible, harsh world.'

I saw him head towards the stables where John was waiting with Paddy, already tacked up and ready to go out on the gallops.

I popped my head round the kitchen door to see how Sue was bearing up under the strain and I couldn't believe my eyes; the place looked like a Chinese laundry! The washing machine and tumble drier were going full tilt and there were two huge cauldrons of vile smelling tail bandages boiling on the stove. Somewhere in the middle was a vibrant, multicoloured pile of all the owners' silks.

'For goodness sake, Sue!' I blurted out. 'If *Ideal Homes* could see this lot they'd have a fit!' I pointed toward the huge pots. 'They had more elegant kitchenware than this in mind when they designed those smart hobs!'

'Don't speak! Don't speak!' wailed Sue, as she gathered up even more laundry to her ample bosom.

It was her way of dealing with the stress, and late last night she'd suddenly decided to give the contents of the tack room a thorough spring clean.

'You know, I think I'd rather have the old kitchen back.' She began to empty the washing machine like a robot. 'When it was shabby, I didn't care what the place looked like, and what's more, it didn't cost me a fortune.'

She grabbed the hot iron. 'I suppose Ben's told you we're going to have to pay most of the ransom ourselves.'

I gasped! Where on earth would Ben and Sue find two hundred thousand pounds? They were already stretched to the limit. Now I noticed there were huge black rings under Sue's eyes.

'We own Red Rag, of course, so we have to find his share anyway.' Sue sighed. 'Tomas's father hasn't got any money and it seems as if our grand Mr Holsborough is strapped for cash right now.' Sue tugged furiously at one of the silks.

'Patsy's father paid up, of course,' she added, 'but that still leaves a huge shortfall.' Sue's voice began to shake. 'My parents have given us their nest-egg and we're re-mortgaging everything else. We'll have to cough up – all the other trainers are, and it'll ruin us!' She slammed the iron down and started to cry.

I'd never seen Sue like this, she was always the strong motherly one, with a quick tongue and a great sense of humour. There wasn't much sign of that now – the blackmailers had seen to that.

When Tomas put his head round the door, he quickly sensed our mood. 'I hope my news will make you better.' He had just come from the phone.

'When my father saw the serum had gone, he started to produce some more. You remember, Becky?' Tomas struggled to find the words. 'It is in the blood of the Kladruby horse! He only has a very little but it is for you and Red Rag, and enough for the horses here. My father will fly here at the end of next week.' Tomas smiled. 'You see, Sue, now you no have to pay!'

I couldn't believe what I was hearing as I ran over to the door. Edward had just arrived to ride Paddy and was as stunned as we were to hear what Tomas had to say. I flung my arms round Tomas's neck.

'Oh, Tomas, that's wonderful news. If only Monkey can hang on a little longer, he'll be OK.'

Tomas looked down, sadly. 'It is some time that Red Rag is sick, Becky. Maybe it is too late for him. Monkey has this fight on his own.'

I had to get out of the kitchen. I just couldn't take any more.

CHAPTER ELEVEN

I recognized Chrissie a mile off. She was wearing a bright red coat and looked every inch a city girl. Her hair was longer and it fell carelessly from under a black hat which she wore with her usual style. She looked great! I ran along the platform and gave her a big hug.

'Well, I could hardly miss you!' I laughed, as I held her back for a better look. 'Although I think you could have slapped on a bit more warpaint, darling; the horses are just going to *love* your outfit!'

'Cheek!' Chrissie stuck her chin in the air in mock indignation. 'I haven't dressed for the animals, you know, and there's no way I'm spending my precious weekend mucking out the horses!' She lifted one black-gloved hand as she eyed me up and down. 'Oh no! I've got plans for you, my country bumpkin – and they don't include the stables!'

I grabbed her bag and led the way to the car. She stopped abruptly as I pointed to the purple one.

'You must be joking, Becky! I can't get in that – I'll clash!'

We fell about laughing and the other travel-weary passengers looked at us as if we were crazy.

We hardly paused for breath on the way back to the Lodge and I felt all the tension and worry of the past few days fall away, as we gossiped and exchanged news on the forty-minute drive home. I gave her the latest on Jamie and the Doncaster trip, but she was just as interested in hearing about Tomas.

'Come on,' she said. 'You're not telling me you don't find this blond hunk just a little bit attractive?'

I had to admit I did enjoy flirting with him, and maybe he did fancy me a bit – but that was all!

'I believe you, Becky. Honest!'

Chrissie was impossible, but she was definitely good for me, especially now!

After supper we went up to my room and listened to music until gradually I started telling her more about the dreadful horse virus and I gave her the latest on the blackmailers. I explained that it would take the international banks several days to transfer the money into the numbered Swiss account, and the trainers were all waiting desperately for the delivery of the serum.

This time, when Chrissie mentioned Tomas, it wasn't in such a light-hearted fashion.

'Look, I know you like the guy, but isn't it a bit of a coincidence that he's over here at this precise moment, *and* his father just happens to have been working on the virus?'

I started to protest. Tomas loved horses, there was no way he would harm them.

'No, wait a minute, Becky. Let me finish. Have you ever wondered why your yard isn't infected? Tomas's precious horse would get it too, wouldn't he?'

I began to feel worried. Chrissie was getting excited now.

'And isn't it just too convenient that the Professor stands to make a bundle and get lots of publicity from developing the serum? Everyone's going to want it after this. The money will come in very handy for the restoration work on his stately home.'

'But he's *helping* us – he's giving us some serum for Red Rag,' I argued.

'Perhaps he feels just a little bit bad that *your* horse has got it,' she conceded. 'If Red Rag had stayed at home, as he was supposed to, he wouldn't have caught the virus. In any case, the titchy bit of serum he's giving you isn't going to make any difference to the rest. He'll make millions from the other trainers.'

I stared at her. I was just beginning to realize what Chrissie was suggesting. She was shouting now.

'Don't you see? It's the perfect cover! "Kind Professor saves Gold Cup Winner", when he's really making a killing.'

'Oh Chrissie, you're right!' I could hardly

believe it. 'They can't lose either way. They'll make a fortune out of the ransom notes *and* they'll rake it in from selling the serum world-wide. The irony is the Professor will be internationally recognised as the saviour of British racing, when he's actually tried to destroy it!'

The more I thought about it, the more it seemed feasible. Tomas had visited some of the yards that had become infected. He was with me when I went to the Knight's on the *Charisma* sales trip, and now Monkey was close to death. I felt myself getting really angry.

'You're right, Chrissie. Tomas had the opportunity and he always seems to be self-conscious about how poor his family is. And there've been times when I've knocked on his door in the evening, just for a chat, and he hasn't been there. Maybe he slipped away with the missing Husak to infect a few more yards.'

Tears came into my eyes and my voice started to shake. 'If I ever find out that it was Tomas who hurt my precious Monkey, Chrissie, I swear I'll kill him!'

'Don't, Becky! Don't get so upset!' Chrissie crouched down beside me and put her arms round me. 'Anyway, killing him would be far too messy!'

I smiled, suddenly feeling strong again and full of determination. My voice sounded firm as I turned to Chrissie.

'We need to get some proof – and we need to get it fast!'

Once Chrissie went back to London, I didn't let Tomas out of my sight, and the moment I had a chance I asked the lads some questions, trying to sound off-hand.

'How's Tomas getting on?' I was alone with Andy in Flytrap's stall, giving him a hand with pulling the mane. 'Do you see a lot of him in the evenings – after work?'

Andy stuttered a bit when he answered. 'I— I don't see a lot of him at night. I think he likes to be a bit— invisible. Not that there's anything wrong with that, of course,' he quickly added. 'But he's not the sort of person you can get close to.' Andy began to look uncomfortable. It was almost as if he was trying to tell me something.

'Do you think Tomas goes out at night, Andy?'

Before he could answer, John swung round the door. 'Are you done? I need some help with the afternoon feeds.' I made a mental note to talk to Andy again as soon as I could.

Tomas carried on as if there was nothing the matter, but I found it hard to be natural with him when we rode out together.

The icy wind penetrated my jacket and I felt strangely exhilarated as Tomas and I approached the start of the gallops. There was an unspoken rivalry between us as we urged our horses forward into a canter. Genesis was a wonderful horse – handsome and strong – and the wintry sun caught his chestnut quarters as he powered up the slope. But for me, Paddy was in a class of his own and as a ride there was no other horse to equal him. There was a passion in his power, a determined quality to the surging motion and, beneath it all, the dangerous feeling he would take any risk, no matter what the consequences.

We weren't allowed to race each other on the gallops but I would have loved to let Paddy go, and I sensed Tomas felt the same! At the end, I stood in my stirrups, completely elated by the thrill of the ride. In spite of myself, I smiled at Tomas, and he at me (those blue eyes got me every time!). I was finding it hard to think of him as a possible killer, and I started to feel flat again as we regrouped to go back for the second lot.

We got the news of the first death on our return. The Governor looked weary and drained as he laid down the phone in the stable yard; there was no expression in his voice as he turned to us.

'The Sullivans have lost Chance Encounter. He died half an hour ago. They think Enigma might be next.'

None of us could take it in. Chance Encounter was their most promising youngster, a half-brother to last year's Grand National winner. I'd seen him just a month ago at Newbury. It was unbelievable that he was dead! And Enigma was the star of their yard. He'd been tipped for this year's Gold Cup. There was no chance of that now.

'The Sullivans have paid up, of course – they're the first – the rest of us will have to find our money now,' Ben said, grimly.

'Why can't they stop this nightmare!' I shouted in desperation into the air. 'It's got to stop! We can't lose all these fantastic horses.' I was sobbing and gulping back the tears. 'They've given everything to racing; it's just not fair that they have to suffer so much! This thing is spreading like wildfire.'

My lips trembled as I asked the question we all wanted to know the answer to. 'W— what about Red Rag? Have you heard from Jessie?'

'He's still alive – just – that's about all I can say.' Ben turned abruptly and made his way back to the house, clearly upset.

I elbowed my way past Tomas and went back to Paddy's stable to get my jacket. As soon as I went in, I knew something was wrong. A quick look confirmed my worst fears – Paddy had a swelling above the joint on his off-fore fetlock. It looked puffy and felt warm, and when I picked up his leg and squeezed it, it was obviously sore.

'Oh great!' I muttered in disgust. 'This is just what we need.'

I imagined Paddy must have knocked his fetlock when he was working or in the stables and, although he wasn't lame, the injury meant that if he ran at the Festival, it could develop into something more serious and put him out of the National. And that was the race we were going for. Ben didn't deserve this, and neither did I.

'We've no choice.' Ben sounded resigned as he finished examining Paddy. 'Edward will be disappointed but we'll have to withdraw Paddy. We can't risk a serious injury at this stage. I'll phone him straight away.'

That night, I spoke to Jamie and told him

that Paddy wouldn't be racing at the Festival.

'Does that mean I won't see you, Becky?' His voice was suddenly anxious. We'd arranged to meet each other at Cheltenham and it was only days away.

'Sorry, you're saddled with me, if you'll excuse the pun. We've got a runner for the Champion Hurdle and the Governor insists I have to go. He's going to be doing a Jamie,' I teased.

'What do you mean, "doing a Jamie"?' he asked, suspiciously.

'You remember Doncaster?' I replied, airily. 'I mean chatting up all the owners and making new contacts – being a slime ball!'

'Hmm!' Jamie was lost for words but I heard him laugh as he said, 'It shows what a clever guy I am – it keeps the money rolling in! You should try it yourself, Becky. You never know, you might shift a bit more of that stinky *Charisma* stuff!'

'Oooo! Thanks a bunch, Jamie.'

Joking over, I began to fill him in on my sleuthing and told him I almost had proof that

Tomas was involved. Jamie didn't really like him anyway and now he was looking forward to riding him into the ground at the Queen Mother Champion Chase.

'Hang on, Jamie. Remember Tomas has a really good horse – Genesis is going brilliantly at the moment,' I reminded him.

'Wait until you see Wild Rover,' Jamie laughed. 'Only the Irish can breed them like this one! It'll be his first outing across the water and that Czech bloke doesn't know what's going to hit him!'

We were both silent for a moment. Suddenly, we were aware that the Cheltenham Festival wouldn't be the same this year. The quarantine restrictions meant a lot of the best horses would be missing, some of them for good. Jamie was sad when I told him there'd been no improvement in Red Rag's condition.

'Jessie actually thinks he's worse,' I found myself saying, though I could hardly bear to admit it. 'The only thing that keeps me going is the thought that even though he was the first to go down with the virus, he hasn't been the first to die. I know it's only a little thing, but it keeps me hoping.'

We made plans to meet up and talk about things in more detail and Jamie promised to keep an eye on Tomas's movements as far as he could. He was almost as angry as I was.

'I don't understand how he can pretend to be a great horse lover when he's actually destroying them,' Jamie paused. 'But one thing's for sure, Pusscat, I'll hunt him down if it is him. It's the least we can do for Red Rag.'

With Jamie's voice ringing in my ears, I stood and looked out of my bedroom window. The wind was whistling through the trees in the orchard and I could hear the measured beat of a shed door banging in the distance. I shivered as I thought of Monkey, alone and in pain. I desperately wanted to be with him.

CHAPTER TWELVE

Andy turned the horsebox into the Cheltenham racecourse and followed the steward's directions to the stable enclosure. I closed my eyes and felt myself transported back one year. I could almost hear the roar of the crowds echoing through the stand, cheering Red Rag to victory in the Gold Cup. That prize wouldn't be ours this time. But the Grand National loomed ahead – that was the one I really wanted this year.

Tomas was sitting up front between Andy and me and was looking around, pretty impressed with the magnificent stand and grounds. We only had two horses entered: Genesis, who was running on the second day, and Brown Sugar, who belonged to Patsy's father. Mr Wilcox wanted his four-year-old to run on the first day, and they'd given Tomas the ride.

Patsy, of course, would be in her element. She'd be dressed to kill and no doubt make the most of her day of glory in the owners' stand, swanning around with the nobs! Andy was helping me with the horses – it was his first time at Cheltenham and he was eager and very impressed. We soon got ourselves organised and settled into our working routine.

The three-day Cheltenham Festival is the major social event of the racing calendar and there's nothing to match the expectant buzz of the crowds. We felt quite elated, even though everyone was talking about the progress of the virus and the fact that the police still hadn't had any success in tracing the blackmailers. I felt quite miserable and was glad to be behind the scenes in the stables.

I wandered down the stalls, looking at the names beside the doors, until I found the one I was looking for.

'Wild Rover!' I exclaimed. 'You could only

be an Irish horse.' I looked him over carefully and had to admit he was quite special. Suddenly two hands came from behind me and covered my eyes.

'Guess who?' said a muffled voice.

I didn't have to, my accelerated pulse rate gave me the answer.

'Jamie!' I breathed. 'You almost gave me a heart attack! What are you doing here?' Jockeys never went near the stables.

'I've come to get you,' he replied, suddenly serious. 'I've been keeping my eye on that Mr Tomas Stravinsky of yours and I've just seen him do another vanishing trick, and sneak away to the back of the stand. I think we should follow him and find out what he's up to.'

As we set off, we tried to make ourselves inconspicuous in the crowds. Sure enough, we spotted Tomas deep in conversation with two sinister-looking strangers. And you know what – it was clear he didn't want to be seen. I managed to get close enough to hear they were talking in a foreign language and Tomas was shifting uncomfortably from one foot to the other, looking around. So this was the proof we were looking for! Tomas was definitely looking suspicious – he *had* to be involved.

Jamie and I sidled away; we decided we'd confront Tomas later. First, there was the race. Tomas and Jamie would be riding against each other, both determined to win. And now Jamie's energy was fuelled by anger. He was out for Tomas's blood.

My emotions were in a turmoil as I led Brown Sugar round the parade ring. He was a young, inexperienced horse from a famous steeplechase line, and he was worth every penny that Mr Wilcox had paid for him. Tomas rode him brilliantly and I had to admit, part of me wanted him to win. The faces of the crowd blurred into each other and I felt as if I was on a macabre merry-go-round that wouldn't stop. Then one particular face came into focus – a flash of red lipstick and a striking black and white coat. It was Patsy of course. She leaned over the railings and yelled at me.

'Tell Tomas I've put my week's wages on Brown Sugar, and he'd better win or I'll see to it he's on permanent mucking-out duty next week!' She winked. 'And that's nothing compared with what Dad will do!' she shrieked with laughter.

Good old Patsy; not exactly Cheltenham behaviour but she made me grin! An Irish lad was leading Wild Rover on the opposite side of the paddock. He was having a job holding the horse, who was prancing about, clearly ready to get going. Rainmaker, trained by the Turners, was the favourite, but I liked the look of the chestnut, Wannabee, an outsider. I'd watched him improve over the season and this might just be his chance.

As the jockeys came into the ring, a lithe mass of colourful silks, they doffed their caps at the owners and trainers and got their last-minute instructions on how to ride the race. I gave Tomas a weak smile as Ben legged him up into the saddle.

He looked down at me. 'Tell me good luck, Becky! I am so happy to be in England and riding at this most famous race!' He looked towards the horizon. 'If only my father could be here to thank you also.'

'Go for it, Tomas!' I found myself encouraging him, aware of a dreadful conflict within myself. 'Ride to win!'

Out of the corner of my eye I saw Jamie, his dark head bent forwards, totally focused on the race ahead.

Andy and I found a good vantage point half-way along the stands and I took a deep breath before I raised my binoculars to study the horses at the start. Some of the handlers were checking the girths and the other horses walked in a circle waiting for the line-up. I spotted Tomas sitting motionless, as if suspended in time, and I panned across to Jamie who was jostling to get into his pole position on the inside rail. He looked as tightly coiled as a spring.

Just then, I saw Jamie stare menacingly at

Tomas. It was as if he'd decided this ride was avenging all the horses which had died, and those like Red Rag who were seriously ill. Then the tape was up, and they were off.

The Challenge Trophy was over hurdles and they were jumped at a fantastic speed. Wannabee went straight into the lead, ahead of Rainmaker and Wild Rover. Tomas had been at the back of the leading pack, but with each fence he was making ground.

'Come on!' I found myself shouting, unsure who I really wanted to win. Then the crowd gasped as Wannabee stumbled badly and appeared to fall across in front of Brown Sugar – but Tomas was OK and, as they approached the grandstand, the roar was deafening. The favourite was being challenged!

Wild Rover and Brown Sugar were making ground on Rainmaker and I could hardly bear to watch as, neck-and-neck, they came to the last hurdle. The three horses were tiring quickly – they'd gone so fast at the start – and I was worried Jamie and Tomas wouldn't have anything left in reserve. Rainmaker's jockey looked desperate as he urged the horse on with the whip. But Wild Rover responded and so did Brown Sugar, jumping the last fence in copy-book style. Now there was nothing to separate the first three horses. Jamie and Tomas were fighting their way to the line.

The crowd roared as Rainmaker, the favourite, just made it and it took a photo finish to separate the other two. Second place went to Jamie. The crowds were thrilled but not the bookies – they never liked it when the favourite won.

As I led Tomas and Brown Sugar into the winners' enclosures I could see the smiles on the faces of the Wilcox family.

'I told you that lad would make the horse go, didn't I, Renee?' Mr Wilcox gave Brown Sugar a hearty slap and reached up to Tomas to slap his thigh in congratulations. 'A good jockey is like a good mechanic, he knows when to give a bit of a tweak here and there – it's instinctive, isn't it, lad?'

Before he could answer, Patsy pushed through the crowds and launched herself at

Tomas, now dismounted.

'Brilliant! Brilliant!' she shrieked, kissing him enthusiastically. 'Mmm, and that wasn't bad, either!' Everyone laughed as Patsy adjusted the brim of her hat. 'You can forget all that mechanic nonsense, Dad. It was the threat of mucking out that did it!'

'Aye, and the promise of a kiss, maybe!' Her father winked at the crowds. He turned to the Governor and lowered his voice. 'Thanks for letting him run, Ben. I know you thought it was a bit soon, but the horse and the lad were hungry – it gave them an edge.'

Mr Wilcox put an arm round his wife and gave her a squeeze. 'Renee and I would like you all to come back to our hotel for a small celebration. And I can assure you it won't be milk we're drinking!' Patsy's Dad was an ex-milkman who'd made his money out of selling second-hand clutches.

Mrs Wilcox was loving every minute of her day in the limelight in the winners' enclosure. She didn't have a clue about horses but she was patting Brown Sugar tentatively and trying to look the part. Suddenly, she held up her hand and wailed. Her new pale blue gloves were ruined, stained and streaked with Brown Sugar's sweat. I giggled, thinking what a good pair she and Claudia would make at the races!

I glimpsed Jamie through the crowds and stuck up my thumbs to congratulate him, as Andy led Brown Sugar on the long walk back to the stables.

Tomas had to take the saddle and weigh in again, and it was some time before I saw him at the lorry, whistling what could have been one of his national songs. When he saw me frowning he stopped, running his hands through his blond hair.

'Becky, you are so sad – why? We finish at number three but you no smile?'

I *was* going to report all my suspicions directly to the Governor and the police, but somehow I still felt quite close to Tomas and I decided I owed it to him to confront him first. I took a deep breath.

'Tomas, you're a great rider and I thought you were a good friend, but—' I gulped back a lump in my throat, 'now I'm sure you want to harm us and it's *you* who's responsible for my Monkey being so ill.'

Tomas opened his mouth to speak, but the thought of Monkey, once so beautiful and now so close to death, made me hurry on, anger and emotion rising in my voice.

'My horse should have been here today but he's lying in a strange stable, fighting for breath – and it's all because of you!' I stabbed at Tomas violently with my finger.

'No! Don't interrupt me!' I was shouting now. 'You had every chance of infecting those yards, you visited half of them with me, and you were certainly very interested in having a good look round!'

Again, Tomas tried to interrupt, but I wouldn't let him.

'No! Let me finish!' I said, firmly. 'You're desperate for money – money to repair your big house, money to buy clothes, money for fast cars, money for good horses... You've been corrupted by lack of money and now you're prepared to kill for it! And don't lie to me,

Tomas! I saw you today! I saw you with my own eyes, plotting with some others out there on the terraces!'

I pointed wildly towards the stand. 'My family and my friends – we took you in and helped you! And this is what you do to us!'

I leaned against the lorry, choking with emotion and sobbing.

I could see Tomas was stunned. He came and stood beside me, saying urgently and quietly, '*Becky*, how can you think this of me? I tell you it is not true!' He took me by the shoulders and turned me to face him.

'Becky, my family is poor now, it is true. But I come from an honourable family – and a family that has always bred horses. I would never, *ever* hurt you or your horses!' He stared at me intently and repeated, 'Becky! You must believe me! I am not wanting to hurt you or your horses! It is the opposite! I am trying to *help* you.'

'But I *saw* you, Tomas. I saw you with those men!' I pulled away from him.

'Becky, those men were from my Embassy.

They are helping your police! The two countries are working – how do you say it? – in unity to catch Husak. They came specially to tell me that my father has the serum ready. He is coming on the plane in three days' time.' Tomas loooked away. 'We are all ashamed that it is a Czech man who has done this very bad thing.'

'But what about all those visits to the yards?' I sounded less convinced now.

'I only went one or two times, with you, Becky. I think you have visited *all* of them, but no one is saying that *you* are a criminal!'

It was true. I had been to all the yards, either on horse business or on my *Charisma* sales trips. Tomas was right. I had jumped to conclusions and deep down I knew there was no way he would harm the horses.

But if Tomas wasn't helping Husak, who was? Time was slipping away. We just had to find out!

CHAPTER THIRTEEN

Tomas had another success on the second day of the Festival, in the prestigious Queen Mother Champion Chase. This time Jamie joined me in the stands and we both cheered until we were hoarse. Tomas was narrowly beaten by the favourite in a really exciting race. I'd caught up with Jamie at teatime on the first day, and told him then that Tomas definitely *wasn't* involved. He took a bit of convincing, but eventually he could see I was right, and he apologised to Tomas in a very sincere way.

Although Jamie didn't have a ride on the second day, I had to work, so we didn't see a lot of each other. But we snatched every spare moment we could to be together. There were only two more days to get through before the

Professor flew in with the serum and, thanks to Tomas's arrangements, we didn't have to pay for it. The other yards were still waiting to hear from the blackmailers that their cheques had been cleared, so we'd be the first yard to get the precious serum.

After dinner I said goodbye to Jamie. I didn't know when I'd be seeing him again so I felt quite low as Ben drove Tomas and I home.

The following day was Gold Cup day so, with our lunch on a tray, we settled down to watch the racing on TV with Sue, in front of a log fire in the study. First, there were the usual interviews and introductory pieces, interspersed with fashion reports – and I must admit, I thought these were a bit of a giggle. Stupidly, I wasn't prepared for what came next, as we heard the commentator's familiar voice:

'And here's a flashback to last year's winner, Red Rag, powering his way to victory with the most promising newcomer of last season – the talented jockey, Jamie Howland.'

There they were – Jamie and Monkey, strong and fit, racing for all they were worth in that final dash across the winning line. Suddenly, I wasn't hungry any more.

'Sorry everyone, I'm going to leave you...' My voice trailed away, as I swallowed down the huge lump in my throat. 'I'll get on with the afternoon stables... We've got an early start tomorrow.' And I knew everyone could tell just how unhappy I was feeling.

The overnight mist was just starting to lift as we turned on to the slip road which led down to the motorway. Ben had lent me his car for the trip to Heathrow to meet Tomas's father, and we soon began to eat up the miles. Then, out of nowhere, we hit a big tailback and we were stuck! We crawled along for what seemed like ages and I soon realized we were in serious trouble.

'Sorry, Tomas, I thought I'd allowed plenty of time for the journey but I didn't expect this!' I was straining to see what was causing the delay. The tailback was at least a mile long.

Tomas looked up from the map. 'Becky, we will miss my father. What can we do?'

'We can't do anything, Tomas. We'll just have to sit tight,' I replied, anxiously tapping my fingers on the side of the steering wheel. Ten minutes later, we passed a jack-knifed lorry which had spilt its load – the cause of all our problems.

'Go, Becky! Go!' shouted Tomas, as we cleared the last of the cones and the motorway opened up in front of us.

I put my foot on the accelerator and then slowed down. 'This is as fast as I can go! I'm not going to push it too hard. It would be just our luck to get caught for speeding!'

Tomas looked anxiously at his watch. 'The plane is here in eight minutes!' He was beginning to panic. 'And we've still got fifteen miles to go!'

I took a deep breath, increasing my speed slightly. 'We'll be late but we should make it before your father clears customs.'

I fumbled for a new CD as Tomas ran his hands through his hair in frustration.

'Don't worry, sport! Keep yer hair on!' I tried joking and smiling at him. 'Planes are always late. We'll be there to meet yer Pa!'

But we weren't. The car park was very full and it took ages to find a space. We were gasping for breath as we burst into the arrivals hall and stood staring at the information board for news of the flight.

'Oh no! It landed twenty minutes ago. Let's head for Arrivals.'

I grabbed Tomas's arm and we shot across the hall. Arrivals looked pretty deserted.

'Maybe they haven't come through yet. We'll give them five more minutes.' But I'd got a sinking feeling in my stomach as the seconds ticked by and there was still no sign of any passengers.

'He's got to be here somewhere!' A note of panic crept into my voice, too, as I scanned the faces of the hundreds of passengers milling around the terminal.

'Let's ask someone,' said Tomas, tensely, edging his way towards an official.

It took us nearly an hour to find out that the

Professor had collected his luggage and cleared customs and, in spite of several tannoy calls, there was no sign of him at all.

'Maybe he take a taxi?' Tomas voiced our last hope as we dejectedly made our way home. But when we finally pulled up on the gravel drive at the yard, we knew something was wrong – and it wasn't long before we found out what.

'They've got him! The blackmailers have got Professor Stravinsky! Quick, they're on the phone!' Patsy ran into the kitchen where we were sipping a cup of tea, dejectedly telling Ben and Sue what had happened. 'They're on the office phone and they want to talk to Tomas.'

We followed as Tomas dashed after Patsy and grabbed the receiver from the desk. This was our first phone contact with the blackmailers and now Tomas was talking rapidly in Czech. After several minutes, he covered the mouthpiece with his hand and turned to the Mainwarings.

'It *is* Husak – he has my father. He says they now have *all* the serum and you must pay money too.' Sue gasped and Tomas quickly went on, 'Everyone must pay, or they will kill him.' Tomas's hand assumed the shape of a gun as he emphasised the threat. His voice was shaking. 'I say I want proof that they have my father, and they go to get him.'

Before we could speak, Tomas held up his hand, straining to listen, then abruptly put the receiver down.

'It is true!' Tomas looked desperate. 'Husak has my father. He has gun at his head. Only blackmailers have the serum now. I am so sorry, Becky.'

Slowly, Tomas turned and walked out of the house. He needed time on his own to take in the latest blow. Moments later, I found him alone in the stable with Genesis.

'I am sure my father was trying to tell me something,' Tomas murmured, as he puzzled over the telephone call.

I grabbed his arm and turned him to look at

me. 'What do you mean, Tomas?'

'My father use very old words – he speak formal Czech.'

My face lit up. 'That's it! He must have been trying to give you a clue!'

'He speak about Husak and the gun and time running out.' Tomas leaned against Genesis, getting comfort from the warmth of his horse and trying to remember. 'He tell me not to be sad, like the little shepherd boy in the Czech story.'

'What story?' I was excited now.

'A story for little children – that sort of story!'

'A nursery rhyme, you mean? Go on, Tomas,' I said gently. 'Tell me the story.'

'I cry when I hear this story as a boy. It tell of a young boy – his mother is dead and the boy is left with his bad stepfather. The boy is cold and hungry and very sad for his mother. He runs away to the mountains and hides in a cave. He is still alone – but no bad man to hit him.'

Poor Tomas had tears in his eyes as he remembered the sad story from his childhood.

'But the boy is happy in the end! An old man find him – a man with a long beard who live alone in another cave.'

'A hermit! And don't tell me,' I raced on, guessing the end of the story, 'the boy helps the old hermit look after his sheep on the mountain, and they live happily ever after.'

'Yes,' Tomas nodded, 'but it was strange that he was trying to tell me that story. I haven't heard it for a long time.'

'Well, it doesn't mean anything to me, Tomas. What would he be trying to tell us, do you think?'

'I don't know, Becky.'

'Well, what would you do in his position?'

'I– I suppose I'd try to tell you where I was.'

'OK, so the story must have something to do with the name of the place where they're holding him. We could try and get some ideas from a map, but where should we start?'

'He landed only three hours ago – he cannot be far. I will get the map from the lorry,' suggested Tomas, 'and see you back in the house.'

When Tomas came back, we spread the map on the kitchen table and started to work our way outwards from Heathrow, but we didn't hold out much hope once we saw just how many towns and villages there were.

It was then that we heard Patsy yell from the study.

'*Yes!* Got it! Why didn't I think of it before?' She came running through to the kitchen. 'Our phone, you know, the new one? It displays the last call. I've just called the number back – it was a call box in Newbury!'

'Well done, Patsy! You're a star!' Now we really had something to go on.

Quickly we leafed through the road atlas until we found the page with Newbury. Seconds later, a name leapt out at me. 'Hermitage! It's got to be Hermitage! Look, Tomas, right here!' Tomas nodded as I pointed to a small village five miles north east of Newbury. 'Remember the hermit in your father's story? Patsy, it's a longshot, and we'll let you know what we find, but phone the police and tell them the call came from

Newbury. The police can trace things from that end.'

Tomas grabbed the map as we ran out to my car.

I could hear Betsy's engine groaning as I coaxed her forward for an extra bit of speed.

'Come on, Betsy! Don't let me down!' I cried, as I swung the car along the winding country lanes.

'Let's look for a post office, and ask there,' I suggested. 'They're bound to know if a stranger has rented somewhere recently.'

Hermitage was a small village with a church, a pub and a pond. We soon found the sub-post office and I parked carefully round the back. Sure enough, they had heard that a man was staying in one of the holiday homes on the Thompson's estate, but they hadn't seen much of him. He'd been keeping himself to himself, they told us.

'It's not far,' said the postmistress. 'If you're

driving, you'll have to go through the estate. But there's a short cut – the public footpath goes right past the door.' She drew us a map.

We left Betsy parked where she was and we followed the footpath which soon led us to the cottage she'd described. It was a converted gamekeeper's house, half-hidden by a wood and with a high wall round the garden at the back. Smoke was coming from the chimney and I shivered, suddenly tense and cold as the late afternoon drew to a chilly end. I hid behind some bushes and watched the front of the house while Tomas disappeared round the side.

A few minutes later he reported back to me. 'There's a way in,' he panted, 'over the back wall.' He looked at me intensely. 'It's high but you and me, Becky, we are fit.'

Without waiting to think it over, I followed Tomas round to the back of the cottage and we began to heave ourselves over the garden wall, slithering down between the ivy which would conceal us while we watched the house. Tomas crept nearer the cottage and, with one swift movement, had climbed part of the way up a tree, to get a better look. There was a light on in one of the upstairs rooms and I jumped as a silhouette crossed the window and snapped the curtains shut.

'Now what?' I whispered, as Tomas crept back to where I was hiding.

'He's got my father up there! I saw him before he pull the curtains – he's tied to a chair!'

Just then, the back door opened and Husak himself stepped out into the garden, slamming the door shut behind him. He put some rubbish in the dustbin and then followed the path round the side of the house. I heard a car door slam and the engine start up. We couldn't believe our luck! He must be driving away!

'Quick, Becky!' Tomas grabbed my hand. 'I don't think he has anyone else to help him here. Now's our chance to get my father out!'

We ran across the garden and tried the back door – but it was locked, and so was the front one. We dashed round the house again, searching for any way in.

'Why don't we break a window?' I whispered hoarsely, looking for a brick.

'Not necessary,' said Tomas, pointing to a small window in what looked like a downstairs loo. 'You are slim, Becky. If I give you a leg up, do you think you could get in? I think it is partly open.'

'Easy!' I replied, eyeing up the small space the window occupied. 'But it's just as well I haven't eaten today!' I had a nasty moment when I was half in and half out of the window, but with a quick wiggle, I'd made it through and clambered inelegantly on to the loo seat below. Seconds later, I was unlocking the back door. Tomas ran in and together we shot upstairs, trying to work out the geography of the rooms.

'He's here!' shouted Tomas, running over to his father who was talking rapidly to him in Czech, stunned to see his son.

'We don't have long – my father think Husak's just gone out to phone.' Tomas was struggling with the knots. 'Downstairs, Becky, in the kitchen,' gasped Tomas. 'Get the serum.

My father say it will be in a plastic box in the fridge.'

I ran down the stairs, nearly falling head-long as I took the steps two at a time. I quickly found the kitchen and spotted an ancient fridge in the corner. Wrenching open the door, I swept the food aside with shaking hands, looking for the precious serum at the back. There they were! Two large plastic boxes and I ripped the lid off the first one – salad! A quick look in the second, and I knew it was the serum. My heart thumping dangerously, I called to Tomas and we fled out into the night, practically dragging the poor Professor with us.

We were well along the footpath, when we heard Husak's car returning. 'Just in time!' I panted, ducking out of the glare of his headlights. 'Thank goodness we didn't drive to the house – we'd have met each other on the road!'

Betsy had never looked so safe and welcoming and I let out a huge gasp of relief as she started first time.'Let's go!'

The Professor flung himself into the back as Tomas struggled with his safety belt in the seat next to me.

'Let's get outta here!' I grinned as I revved the engine.

Back on the main road, we felt we could breathe again as we thankfully blended with the traffic. Tomas turned to his father in the back, smiling and looking relieved, and then peered out of the rear window.

'I don't think he is following us – that was a great piece of driving, Becky!'

Professor Stravinsky sighed, and said something to Tomas in Czech.

'My father says to be careful,' explained Tomas, turning to me. 'Someone is helping Husak – someone English. My father think he is in your family.'

'What!' I gulped, swerving dangerously. 'You– you don't mean Ned?

'No!' Tomas searched for the right words. 'He does not know the name, but thinks it is a person who is close to you and your family.'

Ten miles from home, I screeched

dangerously to a halt – we'd just passed a phone box by the side of the road.

'A phone! We've got to phone the police! If they hurry, they might catch Husak at the cottage!'

We hadn't a moment to lose. The Professor was safe, and now we had the serum – the race to save the horses had begun. And we'd almost run out of time!

CHAPTER FOURTEEN

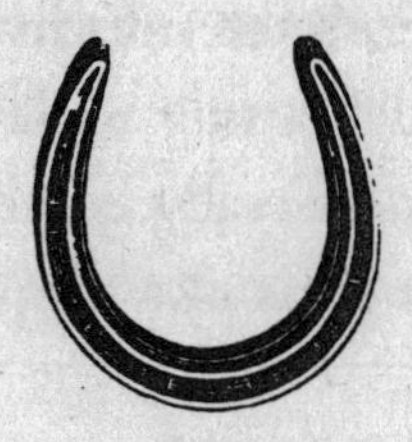

Sue's kitchen had been transformed into Operations HQ. Dick Shannon, our vet, and Ben, were calling out instructions and dividing up the serum for the infected yards. The phone never seemed to stop ringing, as news of what had happened filtered through to the yards and owners. Patsy was busy reassuring them that help was on the way. Our familiar driveway looked unreal as the blue lights of the police cars lit up the bushes like a strobe.

But there was bad news, too: the hideaway cottage was empty when the police had arrived just minutes after our call – Husak had gone. But the Chief Inspector assured us it was only a matter of time before he was caught. We had given a description of his car that we'd seen parked beside the cottage.

Patsy edged her way into the kitchen from the study. 'Here's the information you asked for,' she said, depositing a pile of paper on the table. 'Computers really come into their own when there's a crisis like this,' she announced proudly.

'Well done, Patsy!' Ben peeled off the top sheet. 'OK, John! One of these in an envelope with each batch of serum, then there's no confusion about what dose the trainers have to give.' John handed the first package to Andy.

'That's right!' said Ben, looking up. 'Now tick it off on the list and Go! Go! Go!' He shouted excitedly as Andy ran out of the door and shoved it into the waiting hands of a policeman, who roared off down the drive with it.

'That's the Sullivans and the Dempseys taken care of.' Ben consulted his list. 'Now for the Knights.'

Somehow I found my voice. 'Can I take the serum to the Knights, Ben?'

He looked up and his worried face softened

slightly and a sad look came into his eyes.

'Of course you can, Becky.' He put his hand on my shoulder and propelled me out of the door. 'And good luck!' he called after me, as I climbed into the police car. Dear old Ben, he was such a sweetheart, really.

The siren wailed as we began our dash across the countryside. I tried hard not to let tears form in my eyes as I cradled the package in my lap. And I looked beyond the steamy window and formed a silent prayer.

'Hang on, Monkey! Just one more hour! Please don't give up! Please! Not now!'

Ben had phoned ahead and Jessie was waiting for me as we squealed to a halt at the Knight's yard.

'Becky! Come on, run! Follow me!' she called, as she set off across the main courtyard. 'The Governor and the vet are with Red Rag now!'

I hardly dared look at Monkey as we dashed

into the stable. My once beautiful horse was lying wretchedly on the ground, his sides contorting with his laboured breathing. I squeezed down next to him as the vet rose to fill his syringe, and I gently nursed his head and put my face on his muzzle.

'Oh, Monkey! I'm so glad I'm with you now.' I buried my face in his familiar smell and somehow, from deep inside him, he summoned the strength to move his ears.

Tears were welling in my eyes as I turned to the vet. He was tearing open the swab. 'Not a moment too soon, Becky.' He quickly checked the syringe. 'I must try this, but I can't promise you it'll work.' The vet inserted the needle and then sat back in the straw. 'He's very ill – I'm not sure he'll last the night.'

He patted my back as he made his way to the door. 'Which one next, Gordon?' he asked Mr Knight. And then they were gone.

Jessie put her arms round me and we clung to each other, desperately willing Red Rag to pull through.

It was four in the morning when Jessie came back with two steaming cups of coffee. 'My turn now,' she yawned, and then blinked. 'You're not still talking to him, are you?'

I'd spent the night, wrapped in a blanket, propped up against a wall next to Monkey. Even though he was incredibly weak, he knew I was there and, somehow, I felt my voice was reassuring as he lay close to death. I spoke to him gently, all the familiar words he understood, and sometimes I sang softly as he drifted in and out of a restless sleep. Monkey stirred now as Jessie came over.

'How does he seem to you, Jess?' I asked anxiously, as she put the mugs down in the corner of the stable. She'd nursed him these past two weeks and was sensitive to his every change.

'We'll soon see.' She reached for the thermometer and the stethoscope and listened to his chest. After a second or two she sat back, shaking her head.

'I hardly dare trust my ears.' She listened again. 'His pulse rate seems to have improved.' Jessie studied the thermometer and a small grin crept slowly over her face. 'I don't believe this! Look, Becky! It's down!' She shoved the thermometer into my hand. 'Not by much, but it *is* down!'

Gradually, we allowed ourselves to think Monkey might pull through.

'He's still dangerously hot,' I said, as I felt Monkey. 'Let's see if we can make him more comfortable.'

The immediate danger was heart failure, with his blood pumping furiously round his body in an attempt to cool him down. Jessie and I sponged Monkey down carefully and made sure that the straw around him was clean and dry. When we'd finished, he heaved a huge sigh.

'I think I'll stay a bit longer, if you don't mind, Jessie. I've missed him so much.'

'OK, boss. In that case I'll grab some more shut-eye. I haven't had much sleep lately.'

As I nestled in close to Monkey, I closed my

eyes and tried to imagine the serum working in his bloodstream. It would be like an army of soldiers, coursing through his veins, surrounding and isolating the enemy virus and finally overpowering it. In my head, I willed it on its way, and with teeth clenched, I finally fell asleep.

The vet was back at eight in the morning and I woke suddenly as he put his bag on the floor.

'I see you're keeping him warm,' he said, smiling at the sight of me curled up close to Monkey.

Blushing, I brushed myself down as he began his checks on Monkey. I was frightened to breathe and I noticed his eyebrows arch with surprise.

'I can hardly credit it! This is a remarkably strong horse! You know, I wouldn't have put any money on him coming through the night, but he has, and if I'm not mistaken, I think he's over the worst!'

The relief must have shown on my face, but before I could say anything, he went on.

'Look, Becky, he's by no means out of the wood yet and,' he added kindly, 'I don't want to disappoint you.' He'd seen my face fall. 'But you must understand the damage the virus has done to his lungs.' Dick coughed as if to demonstrate his point. 'If he *does* pull through, it'll take at least six months, or maybe a year, for him to make a full recovery. And I have to warn you the damage could be permanent. His racing days may be over.'

Oh, Monkey! My elation turned to despair, but at least he was still alive!

In the days that followed, the racing world seemed to hold its breath; as if by talking about it, things would go wrong. One or two horses didn't make it, but the good news was the serum seemed to be working. Monkey was still at the Knight's yard – he had a long way to go but, with the quarantine lifted, at least we could visit him now.

There was no news of Husak or the missing

millions, but the police were following several leads abroad. Interpol had found Husak's hire car abandoned at the railway station in Calais. Now he could be anywhere! Slowly, the strain of the past few weeks began to fall away.

'Hey, you guys!' Claudia looked round the door of the study. Ned and I were slumped in front of the TV with Shelley and Scruffy curled up on the floor beside us. 'I know English country life is supposed to be reserved and civilised, but this is ridiculous! If you two get any more laid-back, you're going to stop breathing!'

'What's that, my dear?' Ned was only half listening, and didn't take his eyes off the set.

'I think Claudia's trying to tell us we're a couple of boring old couch potatoes!' I suggested.

'You're darned right I am!' agreed Claudia. 'Do you hear that, Ned? Becky's got it in one!' Claudia pointed the remote control at the set. 'I didn't come all the way back from Florida to be buried alive in rural England! No siree!' She lobbed a cushion in Ned's direction. 'I don't know about you, Major Hunter-Gordon, but there's life in me yet, and I'd like to make a proposal!' Claudia tossed back her long mane of streaked hair. It drove me crazy that she always looked so great.

'American women!' Ned looked at me helplessly. 'In this country it's the men who usually make the proposals but,' he gazed admiringly at Claudia, 'you can proposition me any time you like!' He sat up, suddenly interested.

'Now, Ned, you're not going to wind me up with male chauvinist remarks like that, but I am going to get some action out of you, and it's going to cost!'

There was nothing Claudia enjoyed more than spending money and if she could find someone else's money to spend, then that was just perfect! Ned braced himself.

'Let's go up to town, as you call it. Hit the bright lights, do a bit of shopping in Harrods and around Knightsbridge, and take in one of the West End shows – followed by dinner, of course!'

Claudia didn't believe in doing things by halves and she was just warming up to the idea now.

'Let's make it a "team trip". Tomas can come, and why don't we contact Edward? He's always saying we should call by his London flat for drinks. I think he could well do with a night out after all this. What about Wednesday?'

I thought it was a brilliant idea, but I had one suggestion to make.

'I know one thing Tomas would really love, and that's to go to a football match – and this week there just happens to be a mid-week replay. Could the two of us give the shopping a miss and meet up with you after the game? At Edward's flat, maybe, before we go to the show?'

Claudia laughed. 'It beats me why you think a freezing cold football match rates over an afternoon's shopping. But go ahead, Becky, and ask Tomas. Ned, you fix things with Edward and I'll use this,' she waved a credit card, her beautiful red nails flashing, 'to book the restaurant and the show.'

A few minutes later, there was a note of triumph in her voice as she came back from the phone. 'Well, that's settled!'

There was no point in watching any more TV, we'd missed half the programme anyway, so I decided to take a run over to the yard and ask Ben if he'd give us the day off on Wednesday.

We had no runners that day and Ben said fine, and added that it'd be our last day off before the National, and to go for it. I gave him a big hug, dashed over to the hostel, went through the empty sitting room and knocked on Tomas's door.

I was really disappointed when there was no answer and suddenly I had an awful feeling – a flashback to the other times I'd called round for Tomas and he hadn't been there. I shook those old thoughts out of my mind; the days of spies and traitors were right behind us. Tomas was probably somewhere in the stables doing the evening round, and I decided to scribble him a note instead and ask him what he thought in the morning.

I froze! As I bent down to shove the piece of paper under his door, I heard giggling and the unmistakable sound of Tomas's voice coming along the corridor – from Patsy's room. So that's where he was! And no doubt that's where he'd been all those other times when I'd looked for him. I felt such a fool!

I turned and ran out into the cold night air, bumping slap into the arms of Andy, who was even more surprised than I was.

'Crikey, Becky, you gave me a shock!' The naked bulb outside the cottage clearly showed my distress. 'Are you OK?' He looked concerned.

'I was just looking for Tomas,' I blurted, 'and he wasn't there – well, actually he was!' I sounded so dumb!

'Oh, I see,' said Andy, gently. 'I did try to warn you some time ago, when you asked me what he did in the evenings. I thought you knew.'

'Do you mean he's been going out with Patsy all this time?' I felt really stupid and, before I could stop, I heard myself saying, 'And I thought he was keen on *me*!' I blushed bright red and stumbled past Andy feeling a complete idiot. I just wanted to get away – fast. Thankfully, Ned and Claudia had gone to bed when I got back to the house.

Scruffy jumped up on my bed, snuggling in close. 'Why does life have to be so complicated, Scruffs?' I sighed, as I scratched him behind the ears. 'Tomas is not my boyfriend anyway so I don't know why I mind if he likes Patsy.'

Somehow, I felt cheated that Tomas hadn't told me about him and Patsy and angry that I almost cared!

'Well, from now on I'm going to stick with you, Scruffy. You're a loyal friend, aren't you?'

He cocked one ear and gave me a big lick as we settled down for the night. Another true friend of mine would be coming home at the weekend – the vet had said that Monkey was well enough to travel. That was really something to look forward to!

CHAPTER FIFTEEN

Rumours travel fast and the next day the atmosphere in the yard was icy. Patsy and I didn't speak, and Tomas seemed to be avoiding me. For once there was none of the usual chat as we made our way out to ride and by the time we got there, I was feeling really fed up. When it was my turn to go up the gallops, I decided to go faster than I should, and I could feel myself relaxing as Paddy thundered up the track. He felt wonderful – so powerful and strong – and I felt great as we sailed over the fences. Paddy was built to race, but he had something extra as well. You could sense that he wanted to go just one step further – and that's what made him such a terrifying and thrilling ride.

There weren't many jockeys who could cope with such impulsiveness, but I was mad enough to love it, and even though Edward could be a pain, I had to admire him for sticking with Paddy, too.

'You had your foot down!' Tomas panted, finishing his gallop and waiting for the others to come up.

I had to smile, in spite of myself. His English was getting better by the day, but it still sounded funny with his Czech accent. 'By the way, good luck with Patsy,' I added. 'You kept that a bit of a secret, didn't you?' I thought we might as well get things out in the open.

Tomas turned his head away from me slightly. 'It was stupid of me,' he admitted, 'but I feel bad to like someone else when you are so good to me and teach me so much.' He turned back and I noticed that his blue eyes were glittering. 'It feel like I am a traitor!'

'Well, I did wonder for a while!' I laughed, pleased we could now talk about it.

'It is difficult for me, with owner's daughter,

and Assistant Trainer who is good friend of Governor and his wife. I not know what to do for the best!'

This made us both giggle.

'Spoilt for choice,' I agreed. 'Well, Tomas, you used your good old Czech diplomacy and kept both of us happy, I think!' He looked relieved. 'But it's just as well it's out in the open now,' I joked, 'Jamie was starting to get a bit suspicious, you know!'

'He very lucky!' Tomas was gallant to the end. 'And I am, too!' He added.

'Well, what about Wednesday? Will Patsy mind if you come?'

'No way!' Tomas roared. 'You think *Patsy* like to sit in a big coat and hat in freezing cold, watching a football match?'

'Well, I suppose it isn't *exactly* her scene,' I conceded, smiling. 'She's very Chelsea in her shopping, but *not* in her sporting taste!'

As for me, I'd always been mad about football, ever since Ned took me to my first match when I was eight years old. Now I was a fully paid-up member of the Chelsea

Supporters Club.

'Chelsea play against who on Wednesday?' Tomas knew most of the British teams by heart, and I knew he'd be thrilled with this draw.

'They're at home to Arsenal. Not bad, eh?' I grinned as his face lit up. Between them, the two teams had some of the best players in the country.

'Arsenal and Chelsea!' Tomas was overwhelmed. 'Yes – thank you, Becky. I mean, no please!' He flushed red and swore to himself in Czech. 'You know what I mean.'

I smiled as we settled back into line and headed for home. A day in London would do us all good!

The stands at Stamford Bridge were echoing to the swell of the singing, as the supporters warmed up their voices. First one side, then the other took up their team songs, making a fantastic wave of sound which travelled round the stadium.

We had seats in the West stand which was awash with the blue of Chelsea's home colours. The blocks of Arsenal supporters were picked out in a swathe of red and white, and my skin began to prickle as I let the heady atmosphere flood over me. Tomas could hardly sit still.

'Look, Becky! Over there!' Tomas jumped up, pointing at the new Dutch signing, one of the many famous players running out of the tunnel on to the pitch. The crowds were going crazy. It was close to the end of the season, and the two clubs were within a few points of each other at the top of the Premier League. The strains of *Blue is the Colour, Football is the Game* swelled and then faded as the referee's whistle signalled the start of the game. By half-time we were hoarse. The score was Chelsea two, Arsenal one, but the game could go either way – the teams were so close.

'What about a quick drink and a snack?' I suggested, as we got up to stretch our legs. 'Follow me!'

I grabbed Tomas's arm as we pushed our

way up the stand and started to walk round the long concrete corridor that circled the inside of the stadium.

'What d'you fancy? Is a hamburger OK?'

We stood in line as the queue formed rapidly behind us and when we got back to our seats, the smell of onions hung over the stadium and mingled with the crowds stamping their feet against the cold.

In the directors' box, opposite us on the East Stand, I could just make out a familiar figure – it was Jamie! I was sure of it! What was he doing here when he was supposed to be in Ireland?

He stood up as one of their party squeezed past him to get to her seat. I couldn't believe it! Not Sophie Redmond *again*!

The second half of the match seemed to pass in a blur as I tried to focus on the pitch and not on the couple opposite. The final whistle pierced through me and I mouthed the cheers that were erupting all around. I couldn't even feel pleased about Chelsea's win. Tomas was too wound up by the game to notice something was wrong, and he talked non-stop, reliving every goal and tackle, as we snaked through the crowds and made our way to Edward's mews house in Eaton Place.

Claudia and Ned were already there, drinks in hand, as Edward ushered us into the drawing room.

'Ouch, you feel cold!' said Claudia, as she enveloped me in an expensive-smelling hug.

Edward had said we could use one of his rooms to change for the theatre. After we'd told them the score and a bit about the match, he suggested we go and get ready and then join them for a drink.

'There's a bathroom along the passage, Tomas,' he said, as he indicated the way, 'and Becky, you can use the guest room.'

'That's if you can get in,' said Claudia, drily. 'There are one or two seriously important carrier bags in there!'

'One or two!' snorted Ned. 'More like one

or two hundred, containing some of the most ridiculously expensive garments I have ever seen in my life! I gave up arguing in the end.' He shrugged his shoulders in mock despair.

'But you always tell me I look lovely, darling! Now you know why!' Claudia winked at me. 'But don't be too long, Becky. Edward's laid on the most wonderful champagne.'

I'd decided to wear my Doncaster dress and as I took it out of my overnight bag, Jamie's necklace fell on to the bed. I was going to wear it, but after what I'd seen today, there was no way I'd be seen dead in it, and I threw it back into the bag. I wasn't doing very well in the boyfriend department. Perhaps I should forget about two-faced boys and stick to animals – at least they were always faithful.

Suddenly, and for no real reason, I felt anxious about Monkey, and I got a strong feeling that I had to talk to Jessie, straight away. I pulled on my dress, dragged a comb through my hair, put on some lipstick and went straight to Edward to ask if I could phone.

'I wish she worried about *my* horse half as much as she does about that Monkey creature,' said Edward. 'At least you might pretend to, Becky, while you're drinking my champagne!' We laughed, but I noticed a certain edge to his voice, even though he was attempting to joke.

'Of course you can phone.' Edward took my arm and led me out of the room. 'You can use the one in the study.' He snapped on the light and closed the door behind me.

In contrast to the rest of the house, this room had a very business-like feel, with a solid oak desk and the smell of old leather. This was obviously where Edward did most of his work and there were piles of papers lying next to the computer. As I dialled the Knight's yard, my eyes roamed around the desk and fell on the telephone bill that was lying beside the lamp. I couldn't help noticing that it was huge, but I imagined that a lot of Edward's business deals were done by phone.

I tapped my fingers on the desk, idly scanning the list of numbers as I waited for Jessie to come to the phone. Suddenly, a whole section of them jumped out at me from the page. They all started with the numbers 42-40 – that was Pardubice in the Czech Republic! I knew the code because it was the date of Ned's birthday, the fourth of February, and forty, the age my mother would have been if she'd been alive. That was the way I memorised numbers, and I knew this one because I'd phoned Tomas several times to make arrangements before he came to join us at the yard.

I was really interested in the bill now and I looked at the dates in more detail. Edward had phoned the Czech Republic a lot, and all the calls were since we'd come back from Pardubice. I didn't have time to think about it anymore; Jessie was on the line and had started to reassure me that Monkey was making slow progress and that they'd be coming home next Monday, as arranged.

'There! Do you feel better now?' Claudia could see the relief in my face.

'Right, what about a glass of bubbly before we go to the show?' Edward was in a very expansive mood. He held up the bottle of champagne but before he could pour it, the phone began to ring and Edward was clearly annoyed by the interruption.

'How very inconvenient!' He shoved the bottle into my hand. 'Becky, you'd better make yourself useful – look after everyone while I take this call in the study.'

As soon as he'd gone, Claudia waved her glass of champagne. 'This is lukewarm. You British never chill things properly!' Claudia only drank champagne – and it had to be ice-cold. 'Be a darling, Becky. Run into the kitchen and get some ice. Edward asked you to look after me, didn't he?'

I was smiling as I went into the kitchen. When they married, Ned's shopping bill was going to go up – he wouldn't know what had hit him. It was a good thing that Claudia was wealthy in her own right!

It was an old-fashioned fridge with a small ice box inside. Mr Holsborough was definitely a bachelor – the fridge was full of bottles and not very much food. Then I noticed that smell. I sniffed again. Yes, it was *Charisma*! I'd heard of keeping scent in the *dark* but surely Edward didn't need to keep his scent in the fridge?

Intrigued, I crouched down and peered between the shelves. I couldn't see any *Charisma*, but tucked away at the back was a small plastic box and the smell seemed to be coming from there. I don't know why I looked any further, but I did. Somehow, it seemed odd to put a smart product in a cheap plastic container at the back of a fridge. I was curious as I opened the box.

Inside were dozens of the miniature, *Charisma* sample bottles that I'd given away – but lying beside them, was a small bottle marked simply, AHS-V.

I fell back on my heels, stunned, hardly able to make sense of what I was seeing. Then, as if in a dream, I heard the Professor's urgent voice hissing that last message to Tomas as we ran out of the kidnappers' cottage. Husak had an accomplice! He was someone close to us – almost like part of the family. I couldn't believe it! Surely it wasn't *Edward*?

I jumped as Claudia called, 'Becky, have

you got that ice? We're dying of thirst in here!' Edward was still on the phone as I hurried back to the drawing room. I gave Claudia the ice and pulled Ned to one side.

'Ned, you've got to help me! There's something dreadfully wrong!'

CHAPTER SIXTEEN

I could hear Edward still talking in the study as, urgently, I told Ned what I'd found. 'The African Horse Sickness virus is *here* - in Edward's fridge! It's the same bottle we saw in Professor Stravinsky's lab!' I put up my hand to stop Ned from interrupting me – I didn't have long to explain. 'And there's a huge phone bill in the study and most of its recent calls are to Pardubice. Somehow Edward's involved in all this, I know it – and we've got to find out how!'

Ned slipped out to the kitchen to see for himself, and when he came back he whispered, urgently. 'Leave it to me, Becky, I'll tackle him when he comes off the phone.'

I swallowed hard and prayed that I was

right. Rapidly, I went over the evidence, collecting my thoughts. I was certain Edward *was* the accomplice, but I still wasn't sure how far he was involved.

We heard him replace the receiver and then he came back into the room. Ned gave my hand a squeeze and walked across to Edward who was pouring himself a drink.

'Darned nuisance, people phoning up at this time of night,' said Ned, innocently, just as if he was making conversation. 'By the way, I don't suppose you know if the police are any nearer finding Husak?'

Edward shook his head and said he hadn't heard anything.

'He seemed such a nice chap when we met him at the Professor's lab in Pardubice. In fact, you two seemed to get along rather well. Weren't you going to invest in their research?'

'We did discuss it at the time, but nothing came of it,' Edward replied, adding ice to his whisky. 'You Czech boys still have a bit to learn,' he added, looking at Tomas.

Ned asked the all-important question. 'So,

you haven't been in contact with the Czech lab since we left?'

'No,' replied Edward, beginning to look uneasy at the turn of the conversation. He put down his drink and peered at his watch. 'Curtain's up in half an hour – time we were off, don't you think?'

'I'm afraid we're going to have to give the theatre a miss, Edward,' said Ned, firmly. 'You'd better sit down – I've got something rather important to say.' For a moment, Edward seemed stunned, but he recovered quickly and gave a small, tight laugh. 'What on earth do you mean, Ned?'

'Becky, go and get that phone bill,' said Ned, taking control in a quiet, authoritative way that stopped me shaking.

'Look here, Ned,' blustered Edward, looking at his watch. 'It's just half an hour until curtain's up and it's really time we got going. I can't imagine what you think you're playing at.'

'It's more a matter of what *you've* been playing at,' countered Ned, as I gave him the bill, pointing out all the Czech calls Edward had made.

'You've just told us you haven't been in touch with the laboratory, so what's all this?' Ned handed him the bill.

Tomas moved across from the fireplace and looked over Edward's shoulder. 'That's *our* number!' he said, surprised. 'These are all calls to my father's lab!'

'Perhaps I did make a few calls, I really can't remember. What does it matter?' Edward tossed the bill on to the sofa.

But Ned pressed on, unconcerned. 'I'm afraid the phone bill's just the beginning of it. Becky, will you please bring the box from the fridge.'

The colour drained from Edward's face as I hurried off to get the box. My heart was thumping as I carried it back. Carefully, Ned opened the lid and took out one of the miniature *Charisma* samples.

'We all recognise the *Charisma* samples,' said Ned. 'But what about this bottle?' He was holding up the one marked **AHS-V**.

'It's really none of your business,' said Edward. 'But if you must know, it's a special enzyme we use to enhance the scent.'

'Really? And what does the **AHS** stand for?' Edward couldn't answer. Ned continued, '*I* think it probably stands for African Horse Sickness.'

'It's exactly the same bottle we saw in Professor Stravinsky's lab!' I butted in, unable to keep quiet any longer. 'When I saw it in your fridge, I recognised it straight away! With the virus and the phone bill, we've got all the proof we need about you, Edward!'

Suddenly, he shot up from the sofa and made a run for the door. But Ned was ready for him, deftly extending a foot and tripping Edward, who crashed headlong to the floor. Quickly, Tomas threw himself on to Edward's back, pinning him to the carpet.

'Well done, Tomas,' Ned breathed, quietly. 'Claudia, phone the police and Becky, please lock the door. I don't think Edward will try anything now, but we can't be too careful.'

Ned poured a stiff drink and walked over to

Edward, who was still lying stunned on the floor. 'You can get up now.' Ned helped him to his feet and handed him the drink. 'Come on, you'd better tell us all about it.'

Edward suddenly seemed beaten and old. His head was bowed as, slowly, he started to speak.

'My family lost a lot of money when Lloyds crashed and my father came to me for help. But I'd over-extended myself before the recession and I was in danger of going under, too.'

So Edward was broke! That's why he'd moved his horses to us! I blushed when I remembered how I'd thought it was because we were so famous after Red Rag's Gold Cup win. The truth was, Edward couldn't afford the stable bills at the more expensive Lambourne yards.

'When we went round Professor Stravinsky's lab, it was like an answer to my prayers. I knew people would give anything for the serum to *cure* their horses – all I needed was a plan to make them *sick*. And that took me a while.'

Edward's eyes took on a strange quality, as he remembered how he'd solved the problem. Then suddenly, he swung round to face me, a note of triumph in his voice.

'And that's where you came into it, Becky! Oh, yes,' he added as he saw my look of surprise. '*I* didn't spread the virus. *You* did, my dear! And what a good job you made of it, too! I put the virus into the *Charisma* – and you took it into all the yards! It was in the demonstration spray and the free samples that I encouraged you to hand out to the grooms.'

I shivered as Edward grinned at me, mockingly. 'You were an excellent delivery girl – kindly supplying me with the names and addresses of all the yards you'd visited – and I followed them up with a personalised request.'

Edward paused for a moment, taking a gulp of his drink. He went on. 'I put the Mainwaring's name on the list, as a cover, and it was so convenient when the spotlight fell on Tomas.'

So the source of the lethal virus was the *Charisma!* I was shocked, but also amazed at the simplicity of Edward's plan. Humans were safe – this particular virus only attacked horses – and all Edward had had to do was persuade me to deliver it! Putting the virus in the scent and making it airborne in a spray was a stroke of evil genius; the horses either breathed in the virus when it was sprayed, or it was transmitted from the skin of the grooms, who were wearing it, every time they stroked the horses' heads.

It was simple and effective – and deadly!

'But why didn't our yard get infected, too?' I gasped. I didn't understand.

'I made sure all the Mainwaring girls got an *un*-doctored sample, and your personal bottle, dear Becky, was free of the virus, too.'

For a second his face clouded over. 'What I didn't bank on, was Red Rag having to go over to the Knight's. If he'd stayed at home, he'd have been OK.'

Tears welled up in my eyes as I remembered how Kirsten and Bridget had laughed and squirted each other on my first *Charisma* sales trip – and I'd sprayed the tester in front of Monkey that day! *I'd* filled the stable with the deadly virus and had nearly been the cause of Monkey's death.

'You evil man – and you got *me* to do your dirty work for you!' I felt close to hysteria and found myself attacking Edward, beating him around the head with my fists.

'Becky, Becky! Calm down!' Ned pulled me back, gently. 'He's not worth it! The courts will see he's punished.'

'He can *never* be punished enough for all the suffering he's caused!' I sobbed, as Ned put both his arms round me to steady me.

It was then that the doorbell rang. 'That'll be the police,' said Ned, firmly. 'Claudia, will you be so kind as to let them in.'

It was Chief Inspector Whitmore himself. He listened attentively to everything we had to say and then he formally charged Edward – with extortion and kidnapping. He was also extremely interested in the phone bill.

'As soon as I get back to the station, I'll be

on to BT to trace the most recent calls from this number. If I'm not mistaken, I think we'll find one will lead us to where Husak is hiding.

As Edward was handcuffed and led out of the door by a police constable, the Chief Inspector turned to us, as we stood grouped together looking pale and shocked. A table lamp lay on its side and it was obvious there'd been a struggle.

'Your evening turned out to be more Wild West than West End – but I hope you've not been too disappointed!'

We sank down into the chairs, smiling weakly, feeling shaken but relieved that it was all over and hopeful that Husak would soon be caught. But as Edward was bundled into the waiting car, I realized we had something else to sort out. The Grand National was only a week away and now we had no jockey for Paddy!

CHAPTER SEVENTEEN

The newly-painted fencing was gleaming in the early Spring sunshine, showing off the daffodils which were planted along the length of the Mainwaring's drive. I gulped in the sweet air and turned my face towards the pale sun. I felt both excited and apprehensive about Monkey's return as, anxiously, I scanned the lane looking for the horsebox.

Since Edward's arrest, things had moved quickly. Husak had been caught – traced by the phone bill to Belgium, where some of his relatives lived. Chief Inspector Whitmore had applied for extradition to this country where Husak would stand trial, along with Edward, for the dreadful crime they had committed.

There'd been good news for Ben and Sue.

The Swiss courts had frozen the numbered bank account, and once the blackmailers were convicted, the owners and trainers could make an application for the money to be returned. Hopefully, it wouldn't take too long!

It seemed I was the only one who was disappointed. Although Ben had found Paddy a jockey, an Irish boy who needed a last-minute ride, I didn't feel happy about it. Paddy needed someone really special on board, and I would have given anything to have been able to ride him myself. But an amateur jockey had to be fully accredited, and I didn't have the necessary qualifications. One thing was sure – the jockey would have the ride of his life!

All the talk of Irish jockeys had made me think of Jamie, and suddenly I'd felt really low. He'd phoned, briefly, when he'd heard that Edward had been arrested, but we didn't have much time to talk. Neither of us had mentioned his being at the football match with Sophie, and I'd decided I'd finish with Jamie as soon as the National was over. I didn't want to be two-timed and hurt anymore.

'Here they come, everyone!' I started as John's voice brought me back to the present. 'The Governor's back. Let's meet them in the yard.'

John had spotted the horsebox first, and we spontaneously formed ourselves into a guard of honour, lining up to welcome our favourite horse back home. Jessie was first to climb down from the horsebox and went bright red as we started to cheer.

'Welcome home, Jessie!' I was the first to run up and hug her, and one by one, Patsy, John, Andy and Tomas did the same.

'It's time to unload him,' I said, walking round to the back of the box.

Jessie's face fell a little. 'He's made a lot of progress since you saw him last, Becky, but you'd *all* better prepare yourselves for a bit of a shock.' She pulled back the bolt. 'He's not the same horse who left the yard a month ago.'

I tried hard not to cry as I led Monkey down the ramp. He was weak and unsteady. His coat was dull and his head hung low. I put my arm on his neck, rubbing him gently behind his ears in his favourite place. Slowly, he raised his head, blinking against the bright morning sun, and I could sense he was pleased to be back home.

'Gangway, please!' Slowly, I led Monkey into his familiar stable and as I settled him in, I made him a promise. There'd be no more tears now, I felt quite determined. Monkey needed strong people round him if he was going to recover – and recover he would! I was going to make sure he'd get back to full strength – and maybe even race again.

Monkey would always be my favourite, but I remembered I had another very special horse who also needed my attention. I'd spent six long months improving Paddy's behaviour and performance, and now he was fit for the National. I hadn't met the jockey, but one thing was sure – in three day's time, I'd be handing him the best horse ever. And if luck was with us, they'd win the Grand National!

'Where is he then?' I was angry and accosted Ben the minute he walked into the Aintree stable. I hadn't slept well and I'd been in the stables since dawn. The going was a bit deep and Ben and I had wanted to brief the jockey before he got up in the saddle for the first time.

'If I was Mr up-and-coming Mulligan, about to get the ride of his life on probably the best horse in the country, I'm sure I'd have been here before now, champing at the bit to get on!'

I was pacing up and down, furious and upset.

Ben coughed. 'The jockey's here actually, Becky. I'm sure you'd like to meet him.'

I blushed bright red, cross that I'd been overheard. Ben turned towards the door. 'Let me introduce Paddy's jockey – Mr Jamie Howland!'

I just stood there, my mouth hanging open like an idiot. Jamie! His black hair hung over his forehead in that fantastic way it had, and he gave me his most dazzling smile. He looked unbelievably gorgeous! But what was he doing here? He was supposed to be riding in Ireland. They laughed as I stuttered and stammered, hardly able to speak. 'B— but where's Stephen Mulligan? I— I thought *he* had the ride!'

'Oh, I made that up,' said Ben, mischievously. 'I thought I'd give you a surprise. A pleasant one, I hope!' He winked. 'Goodness knows, you deserve something nice, Becky.'

Jamie came over and put his hands on my shoulders, but I still couldn't believe it!

'Luck was really on our side. When I started to ring around to see who needed a spare ride, I found out that Jamie's horse had been injured,' explained Ben. 'I snapped him up at once. I couldn't believe our luck!'

Neither could I! If there was one jockey who could master Paddy, it was Jamie. Now we stood a real chance!

But this time, I have to say, luck *wasn't* with us, and Paddy didn't win the National. But I wasn't too disappointed – he did everything just right! Jamie rode him like a dream in the most thrilling race, and I was sure Paddy would have won if a loose horse hadn't brought him down near the end. Thankfully, Paddy was unhurt and Jamie only had a few bruises. That was the Grand National. There was a lot of luck involved in winning it and, this time, it just wasn't to be.

We *were* celebrating, though – the Mainwaring yard was in the prize money! Tomas and Genesis had come in third, cheered on by a very proud father, watching them from the stands. The Professor had yelled so hard he'd nearly lost his voice!

Later, when we'd been loading the horses into the lorry, I'd felt strangely alone, even though I'd got dear old Scruffy with me, to keep me company. And my heart had thumped nervously as I'd thought of what I was going to do. I'd promised myself I'd break it off with Jamie, once all this was over, and I frowned as I'd remembered all his lies. I was determined I would do it as soon as possible – but the chance came sooner than I thought.

'Give me a kiss, Pusscat!' I was surprised to see Jamie at the door of the lorry – I thought he'd be celebrating with the others. 'What's the problem, Becks?' he asked. 'Why aren't you happy, too?'

In my head I'd rehearsed the speech a thousand times, and now I told him – *he* was the problem. And by the time I got to Sophie Redmond, I was shouting – not that Scruffy noticed. He was fast asleep, high up in the lorry, nestled into a bale of straw.

'But you've got it all wrong, Becky! I was keeping it as a surprise!' Jamie held my shoulders and looked deep into my eyes. 'I was invited to Chelsea by one of the directors, Richard Sullivan – you know, from Sullivan's the trainers? He wants me to ride for him, and he's offered me a job. I was keeping it a secret, for today.' Jamie looked crestfallen. 'I thought you'd be pleased I'd be living and working nearer you.'

'And what about Miss Redmond? Why was she there?' My voice began to wobble.

'He offered her boyfriend a job, too. Not as good as mine, I should say. They were both there in the box!'

'Oh, Jamie. I've been such a jealous idiot!' I put my arms round his neck and started to cry. 'This is the happiest day of my life!'

'Well, you'd better stop crying, then. Here, let me dry those tears.' And Jamie kissed my face all over.

'OOPS! Caught you!' shrieked Claudia in delight. 'Shall we come back later?' Ned and Claudia were hovering at the door.

'Oh Ned! Claudia! I'm so happy! I thought everything was over, but it's not!'

'So we see! Well, I have a piece of news for you, if you feel up to any more surprises,' said Ned. 'Claudia's bought another horse, Becky, and she's made you co-owner.'

I was completely stunned and I couldn't understand why Claudia had bought me a horse, especially now.

'But I'm so busy, Ned, at work! I don't know if I'll have time to look after one of my own.' I must have sounded *really* grateful!

'I think you'll have time for this one. We did a deal with his owner who's having a bit of a hard time at the moment. In fact, he's going to be out of action for several years to come.' Ned took a deep breath. 'Becky, Claudia has bought Paddy!'

I gasped as Claudia grinned and told me the whole story. 'We went to see Edward in prison, this morning, and I made him an offer he couldn't refuse.'

There went my mouth again – hanging open in surprise. I was hardly able to take it all in and now Claudia was in her element.

'Edward's bound to go bankrupt after all this and I figured Paddy might as well go to a good home, before all the loan sharks muscle in.'

Claudia held out her arms and gave me a huge hug as the truth sank in. 'In any case,' she added, 'I thought it would be a good thing to cement Anglo-American relationships!' And she winked at Ned.

I leaped in the air, grabbing Jamie and spinning him round the horsebox until it began to shake. Paddy was half mine! I could do what I liked with him – even ride him myself, if I wanted! Life was suddenly full of fantastic possibilities and my brain was racing as I tried to take it all in.

The bales of straw began to wobble as we all started to jump and leap around in excitement, and poor old Scruffy was dumped in the middle of the floor, looking extremely surprised. We all felt slightly crazy, I think.

'Oh, dog! Your timing is always impeccable!' We laughed as I hugged him and held him up to Paddy's face. 'Say hello to your new brother! Your family has grown while you've been asleep!'

'I hope I'm included in that,' said Jamie, as he hugged me.

'We'll see!' I grinned. 'I suppose I'll have to put up with seeing a bit more of you now you'll be just down the road!' I felt elated as I gave the final verdict. 'All I can say at the moment, Mr Howland, is that my horse likes you – and that's a pretty good start!'

POSTSCRIPT

I rubbed my eyes and slid gently out of bed, padding across the floor to open the bedroom curtains. I took a deep breath as the sun streamed in through the window and I smiled, pushing the dreadful nightmare to the back of my mind. It had all happened six months ago – although in my dreams I re-lived those awful weeks as if they were just yesterday. I shuddered, reminding myself that the virus had been beaten and life was normal once again. Well, as normal as it ever could be in our crazy yard!

Dick Shannon, the vet, still didn't know if Monkey would race again, but I wasn't giving up hope. And Ben and Sue were making quite a name for themselves with a special feeding and fitness plan they'd just launched.

Tomas had gone back to the Czech Republic and was steadily building on the success he'd had over here. In fact, he was well on the way to becoming one of Eastern

Europe's sporting heroes, according to Patsy, who hears from him just about every week.

Talking of news: Chrissie has passed her first year journalism exams and Ned and Claudia are also celebrating; they've finally set a date for their wedding.

As for me, I've got a new love in the stable and an old love not far away down the road. Life has never been better!

THE END

GINNY ELLIOT first started riding when she was three years old. Her mother used to collect her from school on horseback, leading another horse for Ginny so they could ride back home together.

By the age of seventeen, Ginny had won the Junior European Championships on *Dubonnet*, a horse that had been bought at a Cornish cattle market for £35. This was just the start of many major successes including wins at Burghley, Badminton and both the European and World Championships, on such well-known horses as *Master Craftsman* and *Welton Houdini*. She was also a member of the silver medal-winning Olympic team at Los Angeles in 1984 and Seoul in 1988.

Ginny is currently the Jump Team Coordinator for the British Three Day Event Team and lives in Oxfordshire with her husband, Michael.

Order Form

To order direct from the publishers, just make a list of the titles you want and fill in the form below:

Name

....

Address

....

....

....

Send to: Dept 6, HarperCollins Publishers Ltd, Westerhill Road, Bishopbriggs, Glasgow G64 2QT.

Please enclose a cheque or postal order to the value of the cover price, plus:

UK & BFPO: Add £1.00 for the first book, and 25p per copy for each additional book ordered.

Overseas and Eire: Add £2.95 service charge. Books will be sent by surface mail but quotes for airmail despatch will be given on request.

A 24-hour telephone ordering service is available to Visa and Access card holders: 0141- 772 2281